Lighting Systems

Electrical Installation Series – Advanced Course

E. G. Patterson

Edited by Chris Cox

© 2001, Cengage Learning EMEA

ALL RIGHTS RESERVED. No part of this work covered by the copyright herein may be reproduced, transmitted, stored or used in any form or by any means graphic, electronic, or mechanical, including but not limited to photocopying, recording, scanning, digitizing, taping, Web distribution, information networks, or information storage and retrieval systems, except as permitted under Section 107 or 108 of the 1976 United States Copyright Act, or applicable copyright law of another jurisdiction, without the prior written permission of the publisher.

While the publisher has taken all reasonable care in the preparation of this book, the publisher makes no representation, express or implied, with regard to the accuracy of the information contained in this book and cannot accept any legal responsibility or liability for any errors or omissions from the book or the consequences thereof.

Products and services that are referred to in this book may be either trademarks and/or registered trademarks of their respective owners. The publishers and author/s make no claim to these trademarks.

For product information and technology assistance, contact **emea.info@cengage.com**.
For permission to use material from this text or product, and for permission queries, email **clsuk.permissions@cengage.com**.

British Library Cataloguing-in-Publication Data
A catalogue record for this book is available from the British Library.

ISBN: 978-1-86152-726-4

Cengage Learning EMEA
Cheriton House, North Way, Andover, Hampshire, SP10 5BE, United Kingdom

Cengage Learning products are represented in Canada by Nelson Education Ltd.

For your lifelong learning solutions, visit
www.cengage.co.uk

Purchase your next print book, e-book or e-chapter at
www.CengageBrain.com

Printed by Lightning Source, UK

About this book

"Lighting Systems" is one of a series of books published by Thomson Learning related to Electrical Installation Work. The series may be used to form part of a recognised course or individual books can be used to update knowledge within particular subject areas. A complete list of titles in the series is given below.

Electrical Installation Series

Foundation Course

Starting Work
Procedures
Basic Science and Electronics

Supplementary title:
Practical Requirements and Exercises

Intermediate Course

The Importance of Quality
Stage 1 Design
Intermediate Science and Theory

Supplementary title:
Practical Tasks

Advanced Course

Advanced Science
Stage 2 Design
Electrical Machines
Lighting Systems
Supplying Installations

Acknowledgements

The authors and publishers gratefully acknowledge the following:

Extracts from "BS4533, 1988" are reproduced with the permission of BSI under licence number 2000SK/0371. Complete British standards can be obtained by post from BSI Customer Services, 389 Chiswick High Road, London W4 4AL, United Kingdom (Tel. UK +020 8996 9001).

Extract from CIBSE *Code for interior lighting*, by permission of the Chartered Institution of Building Service Engineers.

Extract from the IEE On-Site Guide reproduced with kind permission of The Institution of Electrical Engineers

Every effort has been made to trace all copyright holders but if any have been inadvertently overlooked, the publishers will be pleased to make the necessary arrangements at the first opportunity.

Study guide

This studybook has been written to enable you to study either in a classroom or in an open or distance learning situation. To ensure that you gain the maximum benefit from the material you will find prompts all the way through that are designed to keep you involved with the subject. If you are studying by yourself the following points may help you.

- ☞ Work out when, and for how long, you can study each week. Complete the table below and from this produce a programme so that you will know approximately when you should complete each chapter. Your tutor may be able to help you with this. It may be necessary to reassess this timetable from time to time according to your situation.
- ☞ Try not to take on too much studying at a time. Limit yourself to between 1 hour and 2 hours and finish with a Try this or the short answer questions (SAQ) at the end of the chapter. When you resume your study go over this same piece of work before you start a new topic.
- ☞ You will find answers to the questions at the back of the book but before you look at the answers check that you have read and understood the question and written the answer you intended.
- ☞ An end test is included so that you can assess your progress.
- ☞ Try this activities are included and you may need to ask colleagues at work or your tutor at college questions about practical aspects of the subject. These are all important and will aid your understanding of the subject.
- ☞ It will be helpful to have available for reference a current copy of BS 7671:1992. At the time of writing this incorporates Amendment No.1, 1994 (AMD8536) and Amendment No. 2, 1997 (AMD 9781).
- ☞ Your safety is of paramount importance. You are expected to adhere at all times to current regulations, recommendations and guidelines for health and safety.

Study times	a.m. from	to	p.m. from	to	Total
Monday					
Tuesday					
Wednesday					
Thursday					
Friday					
Saturday					
Sunday					

Programme	Date to be achieved by
Chapter 1	
Chapter 2	
Chapter 3	
Chapter 4	
End test	

Contents

Electrical Installation Series titles   *iii*
Acknowledgements   *iii*
Study guide   *iv*
Table of contents   *v*

1 Light Sources   1

Part 1 Incandescent lamps   2
GLS Lamp Data   2

Tungsten halogen lamp   3
Tungsten halogen lamp data   4

Part 2 Discharge lamps   4
Basic principles   4
Low pressure mercury vapour lamp and control circuits   4
Low pressure mercury vapour lamp data   4
Switch-start circuit   5
Bi-pin electronic starting   6
Stroboscopic effect   6
Electronic start circuit   7
Electronic starter circuit data   7
High frequency electronic circuits   7
Compact fluorescent lamp   8
High pressure mercury lamp and control circuit   9
High pressure mercury lamp circuit data   9
High pressure metal halide lamp and control circuit   9
High pressure metal halide lamp circuit data   10
Low pressure sodium lamp and control circuit   10
Low pressure sodium lamp circuit data   11
High pressure sodium lamp and control circuit   11
High pressure sodium lamp circuit data   12

Exercises   12

2 Luminaires   15
Definitions and lighting units   15
Light distribution   16
Polar curves   16
Using photometric data   17
    Procedure   17
Experimental data   18
    Procedure: 18
Comparison of polar curves   18
Uniformity of illuminance   19
    Glare   20
Maintenance factor   21
    Lamp replacement   21
Utilisation factor   21
Classification of luminaires   22
Ingress protection (IP)   23
Luminaires in hazardous areas   24
Selecting lamp and luminaire   25
Types of luminaires   25
    Interior luminaires   25
    Recessed luminaires
    Swivel downlight luminaire   25
    Surface mounted decorative luminaire tungsten lamps   26
    Track lighting   26
Fluorescent luminaires   26
    Surface mounted fluorescent luminaire   26
Fluorescent lighting trunking system   26
    High bay luminaire   27
Exterior luminaires   27
    Discharge lamp luminaires   27
    Fluorescent bulkhead luminaire   28
    Tungsten halogen floodlight   28
    Hazardous area luminaires   28

Exercises   29

3 Laws of Illumination   31
Inverse square law   32
Cosine law   33
Standard maintained illuminances for interiors   39
Lighting design   41
    Room index (RI)   41
    Reflection factors of room surfaces   41
Lumen method of calculation   42

Exercises   47

4 Lighting Loads and Special Systems   49
Harmonics   49
    Complex waveforms   50
    Cause and effect of harmonics   50
Lighting loads   51
    Inductive loads   51
    Circuit current   52
    Solution of lighting loads by phasor diagram to scale   55
Balancing loads   57
Extra low voltage tungsten halogen lamp loads   57
    Maximum cable lengths   57
High voltage discharge signs   59
    Fireman's emergency switch   60
    L.V. circuit protection   60
Emergency lighting   61
    United Kingdom laws and standards   61
    European Community requirements   61
    Escape lighting   61
    Design procedure   61
    Levels of illuminance   63
    Open area (anti-panic) emergency lighting   63
    Battery systems   64
    Modes of operation   64

Central battery system	65	
Self-contained system	66	
Choice of systems	66	
Fibre-optic emergency lighting	67	
Standby lighting	67	
Completion of emergency lighting installation	67	
Air handling luminaires	68	

Exercises   **69**

End questions   **71**

Answers   **72**

1

Light Sources

On completion of this chapter you should be able to:

- describe the construction of incandescent and discharge lamps
- explain how light is produced
- draw circuit diagrams for discharge lamps
- describe various methods to overcome stroboscopic effect
- compare the efficacy of different types of lamps
- state suitable applications for each type of lamp

This chapter is in two parts:

Part 1 – Incandescent lamps (Figure 1.1) and Part 2 – Discharge lamps (Figure 1.2).

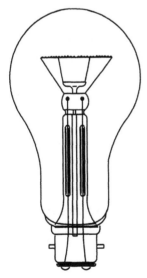

Figure 1.1   Incandescent lamp

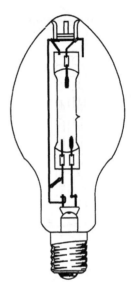

Figure 1.2   Discharge lamp

Lighting plays a significant part in many of our day to day activities and there is a considerable range of lamps available to suit the various activities taking place. Selecting the correct type of lamp is therefore important if it is to meet the needs of a particular activity. For example, large area shop lighting should provide the desired illumination level and colour to help create an ideal environment for employees, to attract customers and promote the items which it is selling. The efficiency of the lighting systems is also a major consideration.

Industrial area lighting must cater for a wide range of activities that range from the production of small electronic components to the manufacture of heavy machinery. Again, the correct choice of lamps is important whatever the activity may be. A satisfactory illumination level, coupled with good colour rendering properties of the lamp, reduces the risk of accidents and errors and helps to create a pleasant working atmosphere which may have a positive effect on productivity and efficiency.

Over the years lamp manufacturers have made considerable progress in the development of more efficient and long-lasting lamps. The aim of this chapter is to provide an outline of the more popular lamps in use, together with their starting methods (where applicable).

Part 1
Incandescent lamps

General lighting service (GLS) lamp

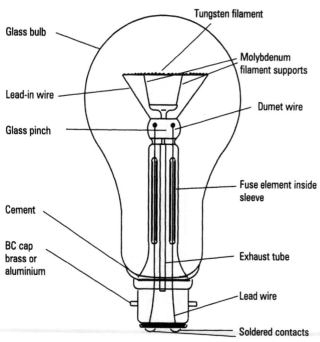

Figure 1.3　　GLS lamp

The construction of the GLS lamp is shown in Figure 1.3 and a variation of this form of light source, known as the "striplight", is shown in Figure 1.4.

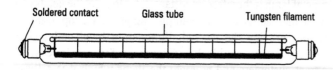

Figure 1.4　　Striplight

The GLS incandescent lamp produces light as a result of the heating effect of an electric current flowing through a tungsten filament wire. Tungsten has a melting point of 3380 °C and if the conditions are correct it can function in the incandescent mode within a few hundred degrees of this temperature. However, where considerable lasting power is required, such as for incandescent lamps that have an average life of 1000 hours, the filament is kept to a much lower operating temperature of around 2500 °C.

Tungsten filaments that are exposed to air at normal operating temperature of 2500 °C will quickly evaporate. For this reason the filament is situated inside a glass envelope (bulb) that has all the oxygen extracted.

VACUUM type incandescent lamps tend to blacken after a long period of operation due to the evaporation of the filament and the consequent deposit of metal particles on the glass wall of the lamp. This loss of tungsten from the filament will eventually cause it to burn out.

Reducing the rate of evaporation of tungsten can be achieved by increasing the pressure on the filament by filling the lamp with an inert gas. Argon and nitrogen are the gases that are generally used. However, the introduction of the gas filling will tend to transfer heat more rapidly from the filament to the lamp wall. If a straight conductor filament was used, the gas would have a distinct cooling effect on the conductor, resulting in reduction of light output.

To overcome this problem the filament is wound into a fine helical coil and the radiation between each turn of the coil raises the overall temperature and improves the light output of the lamp. The efficiency can be increased still further to around 15% when coiled coil filaments are used (Figure 1.5).

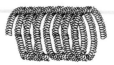

Figure 1.5　　Coiled coil filament

GLS lamp data

- efficacy ranges between 10 and 18 lumens per watt
- good colour rendering
- pearl or clear bulb
- can be operated in any position
- average life 1000 hours for standard GLS lamps
- suitable for dimmer circuits
- no control gear required – connected direct to the source of voltage
- low initial cost
- applications – domestic, commercial and industrial uses
- lamp designation – GLS

Further variations of the incandescent lamp include:
- decorative lamps, for example candle type
- reflector lamps
- special purpose lamps for use in control panels, electric heaters and fires

Try this

Answer the following questions in the space provided:

1. State the reason for using coiled filaments in GLS lamps.

2. Explain how incandescence is produced inside the GLS lamp.

3. Why is tungsten used for the filament?

4. If a 230 V tungsten filament is subjected to a continual reduced voltage of 210 V, what effect will it have on the lamp efficacy?

Tungsten halogen lamp

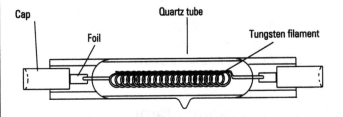

Figure 1.6   Linear tungsten halogen lamp

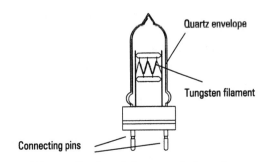

Figure 1.7   Single ended tungsten halogen lamp

The main problem with the conventional incandescent gas-filled lamp is that it loses particles of tungsten from the filament when the lamp has been in service over long periods (that is, lamp blackening). This evaporation process is reduced by enclosing the filament in a small quartz envelope. The envelope can be made smaller because quartz will operate at a much higher temperature than glass, and the increased pressure on the filament slows down the rate of evaporation.

In addition to reducing the dimensions of the lamp, a halogen element is also added to the gas of the lamp. Iodine, chlorine or bromine elements may be used to produce a reversible chemical reaction between tungsten and the halogen. In effect, a regenerative cycle of events occurs where some of the tungsten is evaporated from the filament towards the wall of the lamp and mixes with the halogen to form a metal halide. However, instead of the tungsten depositing itself upon the wall of the lamp, the tungsten halide reverses direction towards the hot filament where they dissociate, causing tungsten to be deposited back on to the filament, while the halogen is available to commence the regenerative process again. To maintain this regenerative cycle a lamp wall temperature above 250 °C is required. Although the tungsten is not always deposited back to the part of the filament it came from, a significant improvement is made in the lumen output and life of the lamp compared to GLS lamps of equivalent wattage. Figures 1.6 and 1.7 show linear and single ended tungsten halogen lamps.

Extra low voltage tungsten halogen lamps combined with precision faced glass reflectors (spotlights) are used extensively for commercial display lighting and domestic applications. The compact 12 volt lamp is supplied from an adjacent 230 V to 12 V transformer or via a remotely positioned transformer. Several lamps can also be supplied from a 12 V track system and a suitably rated transformer.

Tungsten halogen lamp data

- efficacy ranges between 18 and 24 lumens per watt (mains voltage)
- good colour rendering
- operating position – certain low wattage lamps may be operated in any position while linear lamps may have horizontal operation only (reference should be made to the manufacturer)
- suitable for dimmer circuits
- no control gear required – connected directly to the source of voltage
- more expensive than GLS lamps
- precautions –
    care should be taken not to contaminate the lamp surface with greasy fingers, resulting in fine cracks when in operation
    contaminated lamps can be cleaned with a soft cloth and methylated spirit
    under no circumstances must the lamp be touched, either directly or indirectly, when in operation
- average life 2000 or 4000 hours depending on wattage and type
- many applications including – floodlighting, display, security, exhibitions, photographic, general domestic, commercial and industrial uses
- lamp designation – TH

Part 2
Discharge lamps

Basic principles

It was explained in Part 1 that incandescent lamps produce light as a result of the heating effect of current flowing through a tungsten filament. In discharge lamps however, the electric current passes through a gas or metallic vapour and causes it to excite, resulting in the emission of light.

Unlike incandescent lamps, discharge lamps require some form of control gear for their operation.

Low pressure mercury vapour lamp and control circuits

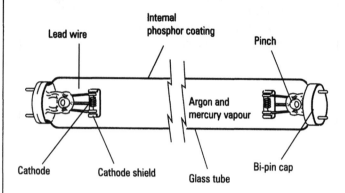

Figure 1.8   Fluorescent lamp

Low pressure mercury vapour lamps are better known as FLUORESCENT LAMPS and they use the effect of phosphorescence to produce light. Two cathode filaments coated with electron emissive material are sealed into a glass tube which contains gases such as argon and krypton with a small quantity of liquid mercury, creating a low pressure region. The inside wall of the lamp is phosphor coated to produce the desired colour of light (for example, white, warm white and coolwhite) (Figure 1.8).

Low pressure mercury vapour lamp data

- efficacy ranges between 38 and 104 lumens per watt
- average life 12 000 hours
- wide range of colour options available
- low operating temperature
- wide range of applications in domestic, commercial and industrial premises
- dimming controls available
- lamp designation – MCF

When the supply voltage is switched on the flow of current causes the cathode filaments to increase in temperature and emit electrons. The negatively charged electrons are then accelerated along the lamp by the high voltage produced by the action of the choke (inductor). The electrical discharge produced generates a considerable portion of ultraviolet energy with a small amount of blue light. By the action of phosphorescence the phosphor coating converts the ultraviolet energy into useful visible light from the lamp wall (Figure 1.9).

Different types of phosphors will emit light at different wavelengths (colours), consequently the mixture used will dictate the colour requirements of the lamp. The introduction of the triphosphor fluorescent lamp, which is internally coated with a mixture of red, green and blue phosphors, has significantly increased the lamp efficacy while having good colour rendering properties.

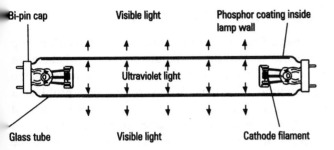

Figure 1.9   Visible light from the fluorescent lamp

Switch-start circuit

Figures 1.10–1.12 shows the switch-start fluorescent lamp circuit.

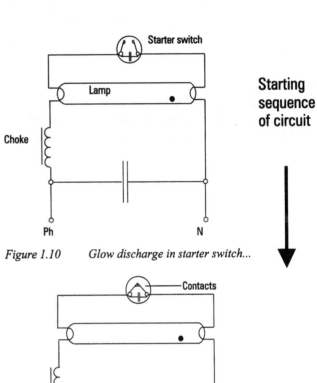

Figure 1.10   Glow discharge in starter switch...

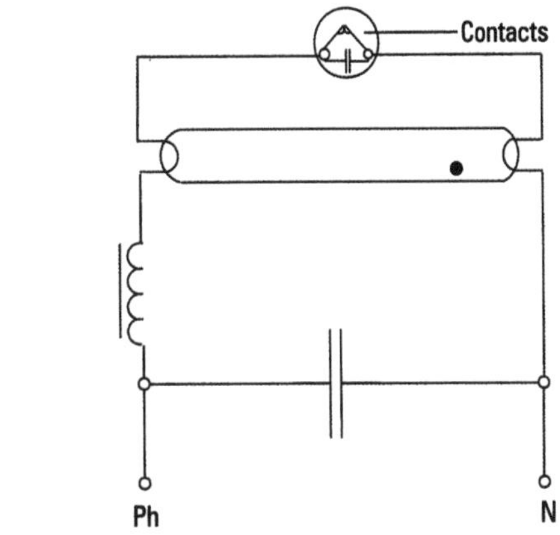

Figure 1.11   ...starter contacts close..

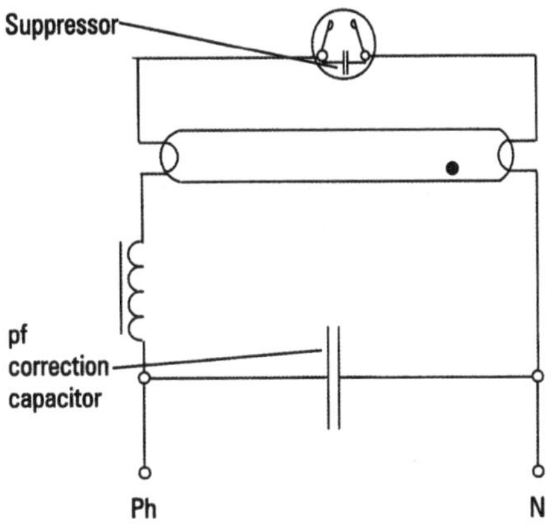

Figure 1.12   ..starter contacts reopen and lamp operates.

Remember

Lamp efficacy is measured in lumens per watt. Light output from a lamp can vary considerably depending on the type and wattage that is selected. The reason for this is because different types of lamps will have a wide variation in ability to convert the electrical power (watts) taken, into light (lumens). Hence, the quantity of light emitted from the lamp divided by the power taken is known as the efficacy of the lamp, measured in lumens per watt.

If there is no control gear (for example choke) required to operate the lamp, input power to the circuit will be the same as to the lamp. However, if control gear is required to operate the lamp (for example fluorescent tube), a power loss will occur in the circuit which will cause the input power to be greater. When calculating installed efficacy of a lighting installation for types of lamps that require control gear for their operation, the power consumption of the control gear should also be taken into consideration in the estimate of luminous efficacy.

Most discharge lamps require a higher voltage than the normal 230 V mains supply to initiate the discharge. The following circuit diagrams illustrate how this is achieved for fluorescent lamps.

A glow type starter switch is connected directly across the fluorescent lamp and contains two bimetallic strip electrodes and an inert gas enclosed in a glass bulb. The starter electrodes are normally open. When the supply voltage is switched on, current flows through the choke (inductor), lamp cathodes, and through the gas in the starter switch. The heat produced by the glow discharge in the starter switch causes the strip electrodes to bend and make contact with each other, allowing a high current to pass through the lamp cathodes. This action heats up the cathode filaments and they emit a large cloud of electrons. Glow discharge in the starter switch ceases and the bimetallic electrodes cool down and reopen. A collapse in the magnetic field across the choke coils occurs which causes a large voltage surge to initiate the discharge between the cathode filaments. The choke now performs another function by limiting the current through the low resistance path inside the lamp.

The function of the capacitor across the starter switch is to suppress any arc across the contacts (radio interference suppression). Due to a choke being incorporated in the circuit a power factor correction capacitor is connected directly across the a.c. supply.

Try this
1. The lamp in a switch-start fluorescent circuit fails to operate (no visible light). State four possible causes for this failure.
   (a)

   (b)

   (c)

   (d)

2. If the power factor correction capacitor became disconnected in the switch-start fluorescent lamp circuit shown in Figure 1.12, would the lamp continue to operate? State the reason for your answer.

Bi-pin electronic starting

The more frequent the operations of the bimetallic strip electrodes in the glow type starter when switching on fluorescent lamps, the greater the possibility of starter failure and limited life-span. To overcome this problem a replacement solid state electronic starter can be plugged directly into the existing glow starter socket in the fluorescent luminaire. It automatically controls the length of time for the preheating of the lamp cathodes prior to starting and it also enables a high voltage pulse to initiate the discharge inside the lamp.

Stroboscopic effect

One of the disadvantages of fluorescent lamps is that they produce a flicker twice every cycle of supply frequency. Although this effect may not be evident to the eye, rotating machinery may appear to be stationary or slowing down. This condition is termed the stroboscopic effect and may cause considerable danger where rotating machines are present. To overcome this effect a lead-lag circuit supplying a twin lamp luminaire may be adopted. One lamp is supplied in the normal manner via an inductor, and the other lamp circuit contains a capacitor in series with an inductor. A 90° phase difference is created in the currents between the two lamps so that one lamp "fills in" the dark intervals left by the other lamp. Figure 1.13 shows the circuit for a lead-lag luminaire.

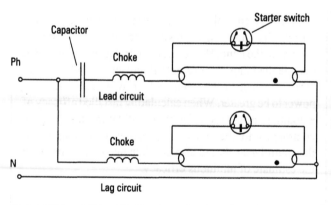

Figure 1.13   Lead-lag circuit

If a 3 phase four-wire supply is available, rows of standard luminaires can be connected to different phases to create a similar effect as the lead-lag circuit (Figure 1.14).

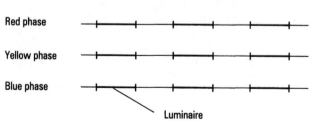

Figure 1.14

Where lathes are used directly under fluorescent lamps, the problem of stroboscopic effect can be reduced by using luminaires containing tungsten filament lamps (normally extra-low voltage) mounted on the lathe. The bright white light is directed onto the workpiece, reducing the possible stroboscopic effect from the general lighting of fluorescent lamps in the area.

Electronic start circuit

The circuit comprises an inductor (ballast) and a capacitor in series with the lamp filaments, making up an R, L and C circuit. An electronic starter is connected in parallel with the lamp. Unlike the switch-start circuit, the electronic starter has no moving parts.

When switching on, the solid-state starter, in conjunction with the other components, provide the cathode current and high voltage to initiate the discharge in the fluorescent lamp (Figure 1.15).

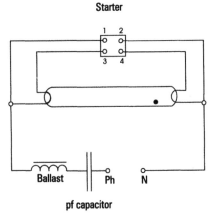

Figure 1.15   Electronic start circuit

Electronic starter circuit data

Electronic starters overcome many of the disadvantages of switch-start circuits, for example:
- flicker-free starting is provided
- extends lamp life
- cuts out supply to failed lamps
- starter built into luminaire and stops pilferage
- cannot be replaced with an incorrect starter

High frequency electronic circuits

Whereas switch-start and electronic starter fluorescent circuits work on normal mains frequency of 50 Hz, high frequency control gear boosts the frequency to 30 000 Hz. This increases the rate of lamp flicker from 100 to 64 000 times per second. Consequently the stroboscopic effect from fluorescent lamps is eliminated. Figures 1.16 and 1.17 show the arrangement for high frequency electronic circuits.

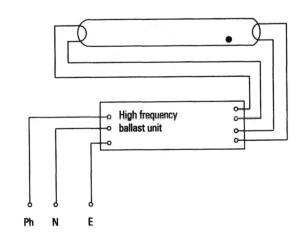

Figure 1.16

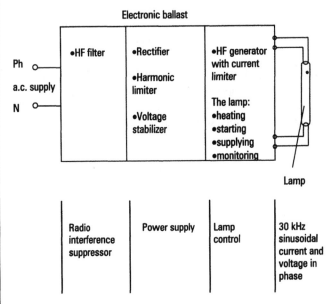

Figure 1.17

High frequency circuit data

Advantages of high frequency circuits over conventional 50 Hz fluorescent circuits include:
- increases lamp efficacy
- no stroboscopic effect (creates flicker-free working environment)
- reduced power consumption in ballasts
- lower operating temperature within luminaire

- almost unity power factor
- reduced running costs
- extended lamp life
- instant, first time starting
- ballast switches off if the lamp fails (stop nuisance flashing on and off of lamp)
- dimming control available
- no problems of noisy chokes, as may be the case in 50 Hz circuits

Compact fluorescent lamp

This type of fluorescent lamp is compact in size to enable it to replace the incandescent lamp and they are often referred to as energy-saving lamps. They use about a fifth of the energy to produce the equivalent light output as the standard GLS incandescent lamp. The life of the lamp is also considerably longer, in the region of 10 000 to 12 000 hours, compared to around 1000 hours for the standard incandescent lamp.

There are many variations of this type of lamp, but they all serve the same purpose by reducing energy bills and lasting longer with increased efficacy (for example a 10 watt coolwhite lamp has an efficacy in the region of 60 lumens per watt).

The type of compact lamp shown in Figure 1.18 is effectively one unit consisting of a tube and a ballast control. If the fluorescent tube or ballast fails, the complete unit has to be replaced. Other types of compact fluorescent lamps have separate tubes to that of the control arrangement.

Energy saving lamps can be used for a wide range of domestic and commercial applications. They are particularly suited where there is the need for prolonged lighting, for example restaurants, corridors, toilets and hallways.

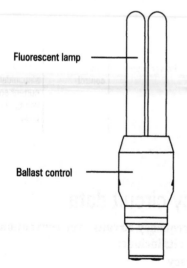

Figure 1.18   Compact fluorescent lamp

> *Try this*
> 1. A factory is to replace all existing switch-start circuit fluorescent luminaires with high frequency electronic control circuits. What effect will this have on the overall power factor, power and tariff costs of the factory lighting load?
>
> 2. Describe how light is produced by a fluorescent lamp.

High pressure mercury lamp and control circuit

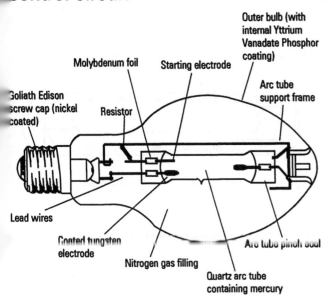

Figure 1.19   High pressure mercury lamp

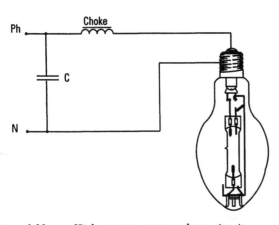

Figure 1.20   High pressure mercury lamp circuit

The inside arc tube of quartz allows more ultraviolet light to be transmitted towards the inner phosphor coating of the outer bulb. This has the effect of increasing the light output and improving colour rendering properties. The outer bulb is filled with nitrogen, or nitrogen/argon gas mixture, and maintains an even lamp temperature. Figure 1.20 shows the circuit for the high pressure mercury lamp.

To commence the discharge, a starting electrode is located close to the main lamp electrode. When the supply is first switched on a small discharge takes place in the argon gas between these two electrodes, the current being limited by the high resistance (10 kΩ to 30 kΩ) in series with the starting electrode. This discharge is now transmitted in the argon gas between the two main electrodes, causing an increase in temperature and mercury vaporisation. Lamp pressure builds up and maximum lamp brilliance is attained when all the mercury is vaporised. The choke limits the current flowing through the lamp and the capacitor is incorporated to improve the power factor.

High pressure mercury lamp circuit data

- lamp efficacy 34–60 lumens per watt
- adequate colour rendering
- lamp takes time to reach its full brilliance
- after switching off, it will not restart until the pressure inside the lamp has fallen
- average lamp life 22 000 hours
- applications – used where colour rendering is not of major importance, for example street lighting, car parks, floodlighting of buildings, general outdoor commercial and industrial uses.
- lamp designation – MBF

High pressure metal halide lamp and control circuit

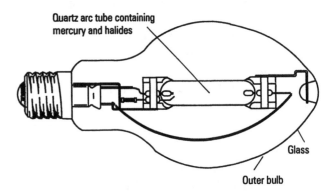

Figure 1.21   High pressure metal halide lamp

Metal halide lamps are similar to standard high pressure mercury lamps but metal halides in the form of thallium, gallium and scandium, are added to the mercury. The inclusion of these halides improves efficacy and colour rendering properties when compared with the standard high pressure mercury lamps (Figure 1.21).

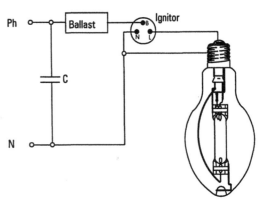

Figure 1.22   Metal halide lamp circuit

One form of starting circuit (Figure 1.22) consists of a switching device (ignitor) and ballast to produce the high starting voltage. When the supply is switched on, the mercury discharge produces a very high temperature and the metal halide additives serve to suppress the mercury spectrum. The result is a crisp white light.

High pressure metal halide lamp circuit data

- lamp efficacy 80–108 lumens per watt
- good colour rendering
- better than MBF lamps
- low power consumption and running costs
- lamp takes time to reach its full brilliance
- after switching off it will not restart until the pressure inside the lamp has fallen
- average lamp life 15 000 hours
- applications – suitable for any indoor or outdoor commercial and industrial uses where good quality lighting is required, for example, stores, exhibitions, sports stadia and TV lighting
- metal halide lamps should be used in enclosed luminaires
- lamp designation – MBI
- the addition of fluorescent phosphor to the inside surface of the glass bulb (MBIF lamps) helps to improve the colour rendering properties of the lamp.

Try this
State the reason for using quartz enclosures in MBI, MBF and TH lamps.

Low pressure sodium lamp and control circuit

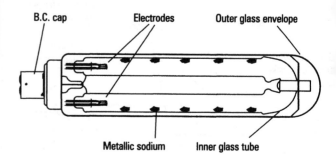

Figure 1.23   Low pressure sodium lamp

Low pressure sodium lamps provide the highest luminous efficacy of all lamps in general use. The construction of the lamp is shown in Figure 1.23.

The inner U shaped glass tube is filled with a neon and argon mixture at low pressure. A small quantity of metallic sodium is also added to the mixture. The outer glass envelope creates a vacuum to keep the inner discharge tube at the optimum operating temperature so that sodium particles can vaporise.

A high voltage is supplied by the control gear to start the lamp. The initial discharge is in the neon and argon gas (glows red). Heat from the discharge gradually vaporises the metallic sodium, causing the discharge to change from red to yellow. A significant reduction in lamp voltage occurs when complete vaporisation of sodium takes place.

Lamp operation is achieved by using a high magnetic leakage autotransformer or a ballast and ignitor. Power factor correction capacitors are used in both circuits (Figures 1.24 and 1.25).

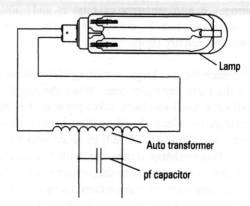

Figure 1.24   Circuit diagram
Lamp operation using a high magnetic leakage autotransformer.

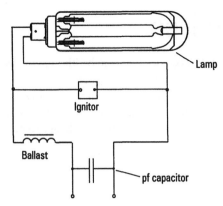

Figure 1.25   Circuit diagram
Lamp operation using a ballast and ignitor.

Low pressure sodium lamp circuit data

- lamp efficacy 70–180 lumens per watt
- poor colour rendering
- low power consumption and running costs
- lamp takes time to reach its full brilliance
- low wattage lamps tend to re-strike fairly quickly after an interruption in supply but higher wattage lamps may take approximately 10 minutes
- average lamp life 16 000 hours
- applications – used where colour rendering is not of major importance, for example, roadways, motorways (good performance in foggy conditions), car parks, floodlighting of buildings, general outdoor commercial and industrial uses
- lamp designation – SOX

High pressure sodium lamp and control circuit

The main disadvantage of the low pressure sodium lamp is its poor colour rendering property, producing a monochromatic yellow light. This characteristic limits its application to outdoor uses. However, if the temperature is increased inside the lamp and therefore the pressure, a much improved colour of warm golden-white is produced. This improved property was made possible by having the inner arc tube made of translucent ceramic material using alumina which is capable of operating at temperatures up to 1500 °C and withstanding hot sodium vapour (Figure 1.26).

The inner arc tube contains sodium and a small amount of mercury and argon or xenon. The outer glass envelope enables the inner arc tube to be maintained at a temperature of not less than 750 °C.

The initial discharge in the argon (or xenon) gas is started by a high voltage pulse of between 2 kV and 4.5 kV. Heat is generated which causes the sodium discharge to commence.

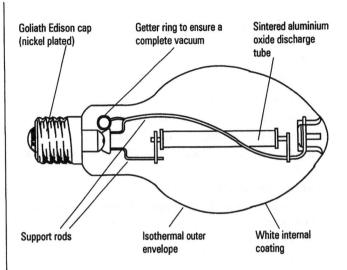

Figure 1.26   High pressure sodium lamp

The addition of mercury causes a higher lamp impedance and an increase in discharge arc voltage between 150 V and 200 V. Maximum sodium vaporisation takes about six minutes, resulting in pleasant golden-white light. Two methods of starting the lamp are shown in Figures 1.27 and 1.28. Power factor correction capacitors are required in each circuit and starting methods may be internal or external to the lamp.

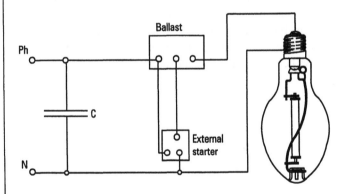

Figure 1.27   SON – starting method circuit diagram

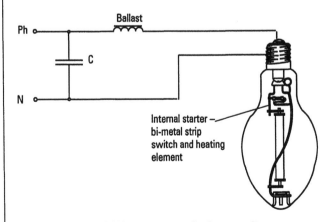

Figure 1.28   SON – starting method circuit diagram

High pressure sodium lamp circuit data

- lamp efficacy 67–139 lumens per watt
- fairly good colour rendering
- low power consumption and running costs
- lamp takes time to reach its full brilliance
- once a hot lamp is switched off, it will not restart until it cools
- average lamp life 24000 hours
- applications – suitable for many indoor and outdoor commercial and industrial uses, for example sports arenas, warehouses, factories and floodlighting of buildings
- lamp designation – SON

Note

Average lamp life

This can vary with different manufacturers.

It must be emphasised that each type of lamp described in this chapter can have a number of variations in construction, wattage, luminous efficacy etc. When detailed information is required on a specific lamp or control system, reference should always be made to the manufacturer's data.

Exercises

1. A small machine shop is to be illuminated by either:
    (a) GLS incandescent lamps or
    (b) MCF low pressure mercury vapour lamps
    Discuss the advantages and disadvantages of (a) and (b) before making a final choice of lamps to be used.

2. (a) Discuss the benefits of using high frequency electronic circuits for fluorescent lamps in commercial and industrial premises.
   (b) A new tungsten halogen lamp is to be inserted inside a luminaire. State the precaution that should be taken when fitting the lamp.

3. During routine maintenance of a factory lighting installation the following faults were observed:
   (a) SOX lamp continued to glow red and outer glass envelope was cracked.
   (b) A fluorescent lamp would only operate when the glow-type starter switch was removed.
   Describe clearly why the lamps in (a) and (b) fail to function correctly. The relevant theory should be used in each case to support your answers.

4. (a) Discuss the reasons why MBI lamps would be more suitable than MBF lamps to illuminate an indoor exhibition area.
   (b) State three advantages of solid state electronic start circuits in comparison to glow-type switch-starters when used with low pressure mercury lamps.

2

Luminaires

Complete the following to remind yourself of some important facts on this subject that you should remember from the previous chapter:

Tungsten halogen lamps must not be _____ when in operation.

The function of the inductor in a switch-start fluorescent lamp circuit is to produce a _____ voltage and limit the _____ through the lamp.

High pressure mercury vapour lamps take time to reach their full _____.

The poor _____ rendering of SOX lamps do not make them suitable for _____ use.

High pressure metal halide lamps are suitable for _____ or _____ commercial and industrial uses.

A capacitor connected across phase and neutral of a SON lamp circuit is for _____ _____ _____.

| On completion of this chapter you should be able to: |

- recognise a range of lighting terms and units
- describe the use of polar curves
- construct a polar curve from a given set of data
- determine the maximum spacing between luminaires
- describe the cause and effect of glare
- explain the reason for maintenance and utilisation factors
- recognise the classification of luminaires
- select luminaires for various situations

Definitions and lighting units

As you progress through the workbook you will need to know a number of lighting terms and units.

LUMINAIRE
an apparatus which controls the distribution of light from the source (lamp or lamps) and includes all the required fixing arrangements, connections and protection.

LUMINOUS INTENSITY
is a measure of the light power of a source in a given direction.

(unit: CANDELA, cd) Symbol I

LUMINANCE
is a measure of the intensity of the light per unit area given off from a surface in a given direction.

(unit: CANDELA PER SQUARE METRE) Symbol L

LUMINOUS FLUX
is the light emitted by a source.

(unit: LUMEN, lm) Symbol Φ or F

LUMINOUS EFFICACY
is the ratio of the luminous flux emitted by a lamp to the power taken by it.

(unit: LUMENS PER WATT, lm/W)

ILLUMINANCE
is the luminous flux density at a surface or working plane.

(unit: LUX, lx) Symbol E

LUMINOUS INTENSITY DISTRIBUTION
is the luminous intensity of a luminaire or lamp in all spatial directions and is normally shown in the form of a polar curve.

Table 2.1

Quantity	Quantity symbol	Unit	Unit symbol
Luminous intensity	I	Candela	cd
Luminous flux	Φ	Lumen	lm
Illuminance	E	Lux	lx
Luminance	L	Candela per square metre	cd/m²
Luminous efficacy		Lumens per watt	lm/W

Light distribution

The purpose of a luminaire is to control the distribution of light from the source (lamp or lamps) in various directions so that the general level of illuminance (LUX) is satisfactory to personnel and the environment in which activities are taking place.

Polar curves

A polar curve is a schematic figure of the luminous intensity (CANDELA) distribution of the luminaire. The shape of the polar curve indicates the way in which the luminaire controls the light distribution from the lamp. Figure 2.2 illustrates the polar curve from an industrial type of luminaire shown in Figure 2.1

Figure 2.1   Luminaire

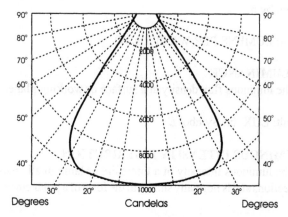

Figure 2.2   Polar curve for a high bay reflector with MBF lamp

Axially-symmetrical luminaires produce a uniform intensity distribution as shown in Figure 2.2. Small luminaires containing incandescent GLS lamps produce similar polar curves.

The linear luminaire for fluorescent lamps (Figure 2.3) does not have an axially-symmetrical distribution. At least two polar curves are required – one on the short axis and the other on the long axis. Photometric data from manufacturers' catalogues make reference to polar curves that represent axial A and transverse T luminous intensity distributions (Figure 2.4).

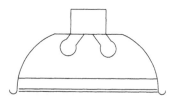

Figure 2.3   Fluorescent luminaire

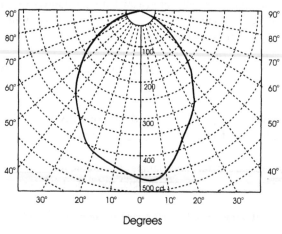

Figure 2.4   Polar curve for fluorescent luminaire

Polar curves for various types of luminaires can be obtained from the photometric data issued by the manufacturer. They can also be plotted: (a) from given set of data (Figures 2.5 and 2.6) or (b) obtained by experiment (Figure 2.7).

2

Luminaires

Complete the following to remind yourself of some important facts on this subject that you should remember from the previous chapter:

Tungsten halogen lamps must not be _____ when in operation.

The function of the inductor in a switch-start fluorescent lamp circuit is to produce a _____ voltage and limit the _____ through the lamp.

High pressure mercury vapour lamps take time to reach their full _____.

The poor _____ rendering of SOX lamps do not make them suitable for _____ use.

High pressure metal halide lamps are suitable for _____ or _____ commercial and industrial uses.

A capacitor connected across phase and neutral of a SON lamp circuit is for _____ _____ _____.

On completion of this chapter you should be able to:

- recognise a range of lighting terms and units
- describe the use of polar curves
- construct a polar curve from a given set of data
- determine the maximum spacing between luminaires
- describe the cause and effect of glare
- explain the reason for maintenance and utilisation factors
- recognise the classification of luminaires
- select luminaires for various situations

Definitions and lighting units

As you progress through the workbook you will need to know a number of lighting terms and units.

LUMINAIRE
an apparatus which controls the distribution of light from the source (lamp or lamps) and includes all the required fixing arrangements, connections and protection.

LUMINOUS INTENSITY
is a measure of the light power of a source in a given direction.

(unit: CANDELA, cd) Symbol I

LUMINANCE
is a measure of the intensity of the light per unit area given off from a surface in a given direction.

(unit: CANDELA PER SQUARE METRE) Symbol L

LUMINOUS FLUX
is the light emitted by a source.

(unit: LUMEN, lm) Symbol Φ or F

LUMINOUS EFFICACY
is the ratio of the luminous flux emitted by a lamp to the power taken by it.

(unit: LUMENS PER WATT, lm/W)

ILLUMINANCE
is the luminous flux density at a surface or working plane.

(unit: LUX, lx) Symbol E

LUMINOUS INTENSITY DISTRIBUTION
is the luminous intensity of a luminaire or lamp in all spatial directions and is normally shown in the form of a polar curve.

Table 2.1

Quantity	Quantity symbol	Unit	Unit symbol
Luminous intensity	I	Candela	cd
Luminous flux	Φ	Lumen	lm
Illuminance	E	Lux	lx
Luminance	L	Candela per square metre	cd/m^2
Luminous efficacy		Lumens per watt	lm/W

Light distribution

The purpose of a luminaire is to control the distribution of light from the source (lamp or lamps) in various directions so that the general level of illuminance (LUX) is satisfactory to personnel and the environment in which activities are taking place.

Polar curves

A polar curve is a schematic figure of the luminous intensity (CANDELA) distribution of the luminaire. The shape of the polar curve indicates the way in which the luminaire controls the light distribution from the lamp. Figure 2.2 illustrates the polar curve from an industrial type of luminaire shown in Figure 2.1

Figure 2.1   Luminaire

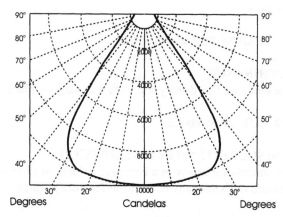

Figure 2.2   Polar curve for a high bay reflector with MBF lamp

Axially-symmetrical luminaires produce a uniform intensity distribution as shown in Figure 2.2. Small luminaires containing incandescent GLS lamps produce similar polar curves.

The linear luminaire for fluorescent lamps (Figure 2.3) does not have an axially-symmetrical distribution. At least two polar curves are required – one on the short axis and the other on the long axis. Photometric data from manufacturers' catalogues make reference to polar curves that represent axial A and transverse T luminous intensity distributions (Figure 2.4).

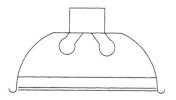

Figure 2.3   Fluorescent luminaire

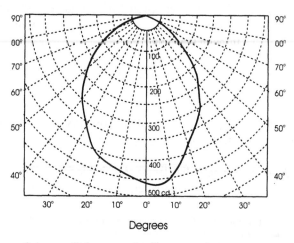

Figure 2.4   Polar curve for fluorescent luminaire

Polar curves for various types of luminaires can be obtained from the photometric data issued by the manufacturer. They can also be plotted: (a) from given set of data (Figures 2.5 and 2.6) or (b) obtained by experiment (Figure 2.7).

Using photometric data

The polar curve data for a symmetrical luminaire containing a tungsten filament lamp is as follows:

Angle from the vertical plane in degrees	0°	10°	20°	30°	40°	50°	60°	70°	80°	90°
Luminous intensity in candelas (cd)	450	470	475	470	420	325	240	175	0	0

The curve can be constructed on polar graph paper or on a set of equally spaced semicircles (or circles if above 90°) drawn on lined paper. Space between each line indicates the value of luminous intensity in candelas. Lined paper is used in this example (Figures 2.5 and 2.6).

Procedure

(a) Construct a vertical line in the centre of the lined paper and from the top of the line produce sufficient semicircles of equal spaces for the values of luminous intensity (cd) (Figure 2.5).

(b) From the top of the centre line produce a set of lines at 10° intervals from the vertical as shown.

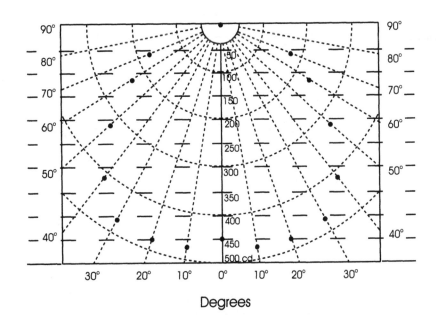

Figure 2.5   Plotted points for polar intensity curve

(c) Plot the values of luminous intensity (cd) at their respective angles shown on graph. Plot on both sides of the vertical plane.

(d) Sketch the polar curve between the points indicated (Figure 2.6).

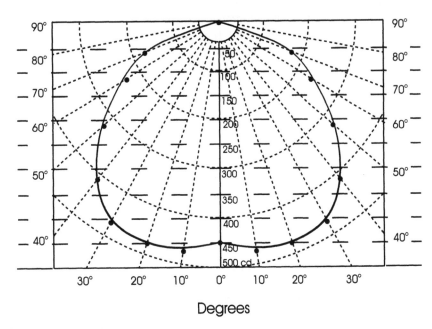

Figure 2.6   Polar intensity curve

Experimental data

A luminaire is mounted in a position so that various readings of photometric data can be taken. The walls of the room should be painted with a matt black finish and no other source of illumination should be present in the room.

Procedure:

1. One wall (or board) of the room is marked at 10° intervals, up to or above 90° as shown in Figure 2.7 (larger intervals, for example 15° or 20°, can be used).

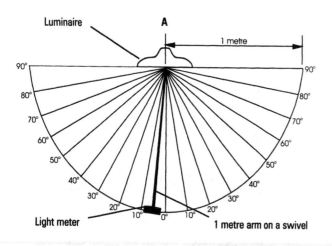

Figure 2.7

2. The luminaire under test is mounted at Position A and a one metre length arm which is able to swivel is also pivoted at this position.
3. A light meter measuring the illuminance of the source is fitted to the bottom end of the arm.
4. The swivel arm is then turned through 90° (or above, depending on the type of luminaire). Illuminance values in LUX are then taken at 10° intervals.
5. Using the formula

$$E = \frac{I}{d^2}$$ (to be considered later in the text)

where E = illuminance in LUX
       I = luminous intensity in CANDELA
       d = distance from source in metres

By using a one metre distance from the luminaire the calculation for the luminous intensity I is simplified.

For example $E \times d^2 = I$

A table showing a set of results is then compiled.

The results of the experiment are then plotted in the same manner as in Figures 2.5 and 2.6.

Figure 2.7 illustrates the various positions of the light meter.

The example shown in Figure 2.7 is for an axially symmetrical luminaire. However, certain luminaires such as fluorescent types do not have axially-symmetrical intensity distributions. Consequently it is common practice to produce fluorescent luminaire polar curves in the vertical plane containing the axis of the luminaire (axial polar curve) and in the vertical plane at right angles to this (transverse polar curve).

> *Try this*
> From the polar intensity curve in Figure 2.6 determine the luminous intensity at 35°, 55° and 65°.
>
> 35° = _____
>
> 55° = _____
>
> 65° = _____

Comparison of polar curves

Figure 2.8 shows the polar curves from two different types of luminaires plotted on the same polar graph paper. Curve A gives the impression that this luminaire emits a lot more luminous flux than luminaire B. However, this is not so because the solid angle subtended by the beam of lights in curve A is less than the solid angle subtended by curve B. Although luminaire A has a higher intensity of downward luminous flux its spread of luminous flux is, however, much less than that from luminaire B.

Careful consideration must be given to the spread of luminous flux when selecting and positioning luminaires.

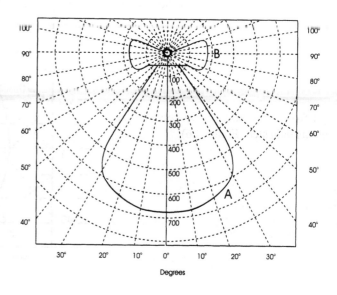

Figure 2.8 Polar curves for two axially-symmetrical luminaires

Uniformity of illuminance

The uniformity of illuminance for an indoor lighting scheme is one of the many important factors that must be considered during the initial planning stage. Uniformity of illuminance is achieved by limiting the spacing between the centres of each luminaire.

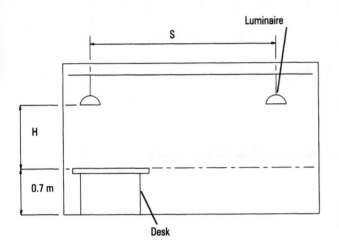

Figure 2.9 Maximum spacing between luminaires

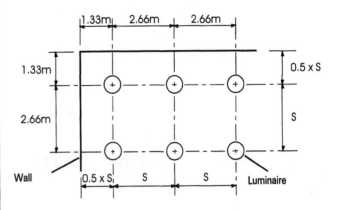

Figure 2.10 Plan view of lighting layout showing maximum spacing

The maximum spacing (S) permitted is determined by the luminous intensity distribution (polar curve) of the luminaire and its mounting height (H) above the working plane (for example desk height above floor, 0.7 m in offices or 0.85 m above floor in industry) (Figure 2.9).

Spacing to height ratio is the spacing between the centres of the luminaires divided by their height above the working plane.

Maximum spacing information for symmetrical luminaires may be shown in the photometric data as SHR MAX, meaning space-height ratio maximum. For example, if a SHR MAX 1.4 is stated for the luminaire in Figure 2.9 and the mounting height of luminaire above the working plane is 1.9 m then the maximum spacing on either direction can be calculated as follows:

$$\frac{S}{H} = \frac{S}{1.9} = 1.4$$

Therefore, maximum spacing S = 1.9 × 1.4 = 2.66 m (Figure 2.10).

If the spacing-height ratio is exceeded then there will be areas between luminaires which will have a serious reduction of illuminance (Figure 2.11).

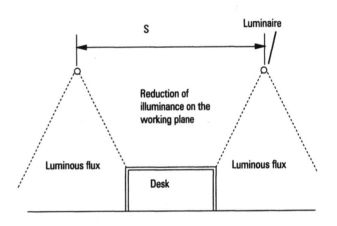

Figure 2.11 Incorrect spacing of luminaires

It is recommended that the ratio of the minimum illuminance to the average illuminance over the working plane (or task area) should not be less than 80 per cent.

In the case of fluorescent luminaires that do not have an axially-symmetrical intensity distribution, maximum spacing information stated in the photometric data may indicate:

SHR MAX and SHR MAX TR

In these circumstances, three conditions must be complied with:
1. The spacing in the transverse direction (SHR TR) must not exceed SHR MAX TR stated.
2. The spacing in the axial direction (SHR AX) must not exceed the SHR MAX stated.

3. The actual spacings in the two directions (SHR AXIAL and SHR TRANSVERSE) when multiplied together must not exceed (SHR MAX)^2.

Figure 2.12 illustrates a plan view of both axial (AX) and transverse (TR) directions for fluorescent luminaires.

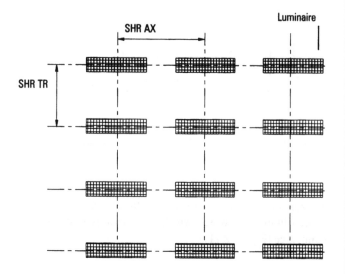

Figure 2.12   Plan view of axial and transverse directions for fluorescent luminaires

It must be emphasised that the spacing to height ratios stated will only give guidance for unobstructed areas. In practical situations many rooms are obstructed and shadows from, for example, high standing screens, cupboards and filing cabinets may cause a problem. In this situation the actual spacing to height ratios must be reduced accordingly. One simple rule that may be applied to the general lighting of an area is that it should be designed so that no activity relies upon the illumination from a single direction.

Try this

The maximum space-height ratio for a symmetrical luminaire is 1.3 and the height of the luminaire above the working plane is 2.1 metres.

(a) Determine the maximum spacing between luminaires.

(b) What would be the effect on the level of illuminance at the working plane if the spacing between the luminaires was exceeded?

Glare

We experience glare when bright objects (such as luminaires) are excessively bright in relation to the general surroundings.

Glare from a lighting installation can make working conditions very unpleasant, giving rise to headaches, general fatigue and a reduction in output from individuals. This is known as "discomfort glare". The degree of discomfort glare is represented by a GLARE INDEX, which may be calculated for the conditions of the installation concerned.

Disability glare is glare which makes it very difficult to see detail. This can occur where a bare lamp is seen in close relation to the work; for example trying to read a meter placed close to a bright light that is in the direct line of vision.

Figure 2.13 illustrates the GLARE ZONE within which the luminance of a luminaire should be restricted so as to avoid DIRECT GLARE. This form of glare is most critical when high intensity light from a luminaire is emitted at an angle above 60 degrees from the vertical.

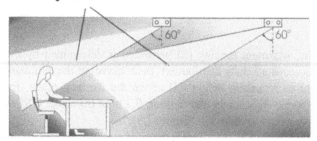

Figure 2.13   Direct glare zone

INDIRECT GLARE has a critical zone within approximately 20 degrees from the vertical, as illustrated in Figure 2.14.

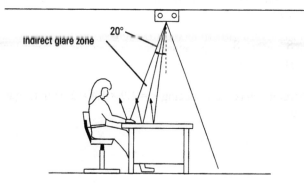

Figure 2.14   Indirect glare zone

It would be a difficult task to work under these conditions where the image of a high intensity source of light is reflected into the eyes from a glossy surface.

Maintenance factor

During the initial design stages for an interior lighting scheme the designer must make allowances for the depreciation of illuminance. In the past, the value of maintenance factor used in the design calculations only took account of the losses of illuminance due to dirt accumulating on lamps, luminaires and room surfaces. The new definition of maintenance factor is, "the ratio of maintained illuminance to initial illuminance", and it takes into account all losses that occur, including lumen maintenance and lamp failure losses. These additional items enable a more accurate valuation of the maintenance factor to be made.

The maintenance factor (MF) is a product of several factors, as follows;

$$MF = LLMF \times LSF \times LMF \times RSMF$$

where
- LLMF is the lamp lumen maintenance factor
- LSF is the lamp survival factor (used only for group lamp replacement programmes)
- LMF is the luminaire maintenance factor
- RSMF is the room surface maintenance factor

Both LLMF and LSF will vary depending on the type of lamp installed.
LMF will also vary depending on the type (construction) of luminaire, location and frequency of cleaning.
RSMF is governed by the frequency of room surface cleaning (effect on surface reflectance).

The graph in Figure 2.15 illustrates the reduction of light output over a period of months where luminaires have not been cleaned.

Note:
Reference must be made to manufacturers' data in order to calculate MF. The values of MF stated in Chapter 3 and in other parts of the text will assume that the calculation has taken place.

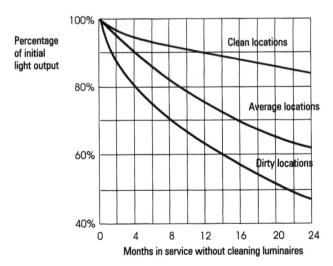

Figure 2.15

Lamp replacement

Light output from a lamp diminishes during the course of time, the rate of depreciation depending on the type of lamp. Tungsten filament type lamps usually fail before any significant decline in output occurs. Nevertheless, there is some loss of light output as the lamp ages. Other forms of light sources such as fluorescent and various types of discharge lamps show a fairly significant reduction in light output before they fail. The probability of lamp failure and the decline in light output will dictate the lamp replacement time.

In certain situations it may be desirable to replace all of the lamps at planned intervals, even if they are still operating, because the light output has dropped to such a level at which it is more economic to replace lamps. This method, known as group replacement, ensures that the level of illuminance is maintained throughout the area. Group replacement will be much cheaper if it can be planned during a period in which the luminaires are cleaned. Where a decision is made for individual lamp replacement only, failed lamps should always be replaced as soon as possible.

Maintaining the correct level of indoor illuminance is important for many reasons, depending on the situations, for example:
- showrooms (for example furniture and clothes) need to highlight their commodities in the best way possible to attract the eye of the customer
- a well-lit factory contributes to a safer working environment
- general lighting in offices should not fall to a level which may cause discomfort and reduce performance of personnel
- the correct lighting design layout and choice of luminaire are particularly important where several VDUs are in use (to minimise reflection and glare)
- the general level of illuminance of hospital ward lighting must not cause any discomfort to patients, and reading (or examination) lighting positioned above each bed should be a type of luminaire that confines its beam of light over a narrow angle so that patients in adjacent beds are not disturbed
- local lighting in operating theatres must provide a very high level of illuminance upon the operating table.

Utilisation factor

The utilisation factor is the proportion of the luminous flux which reaches the working plane. It is a measure of the efficiency with which light emitted from the lamp is used to illuminate the working plane.

$$\text{utilisation factor} = \frac{\text{luminous flux reaching working plane}}{\text{total luminous flux emitted from luminaire}}$$

The UF allows for direct and indirect illuminance falling on the working plane. Consideration is given to room proportions, reflectances of room surfaces (CEILINGS, WALLS, FLOORS) and light distribution of luminaire.

For each type of luminaire a range of utilisation factors can be obtained by making reference to the photometric data supplied by the manufacturer. The UF values are then used in the design calculations for a lighting scheme. Use of this data will be discussed in Chapter 3.

Try this

1. State the difference between disability glare and discomfort glare.

2. What is the purpose of using a maintenance factor during the initial design stages of a lighting installation?

Remember
General lighting provides a fairly uniform illuminance on the working plane over the entire area of an interior.

Localised lighting is designed to illuminate an interior and also provide higher levels of illuminance at different parts of the interior (for example work areas).

Local lighting provides illumination over a very small area which is occupied by the task (for example table light and lathe machine light).

Classification of luminaires

In addition to its normal function of directing light to a surface, a luminaire must also be safe during operation and be able to withstand the prevailing conditions in which it is installed.

British Standard Code of Practice BS 4533 (EN 60598) is applicable to most luminaires. However it does not apply to luminaires that are intended to be installed in potentially explosive atmospheres (hazardous areas).

Table 2.2 shows a brief extract from the section of the Code of Practice that classifies luminaires according to the type of protection provided against electric shock.

For the complete BS requirements, reference should be made to the current Code of Practice.

Table 2.2

Class	I	II
Type of protection	A luminaire in which protection against electric shock does not rely on basic insulation only, but which includes an additional safety precaution in such a way that means are provided for the connection of accessible conductive parts to the protective (earthing) conductor in the fixed wiring of the installation in such a way that the accessible conductive parts cannot become live in the event of a failure of the basic insulation.	A luminaire in which protection against electric shock does not rely on basic insulation only, but in which additional safety precautions such as double insulation or reinforced insulation are provided, there being no provision for protective earthing or reliance upon installation conditions.
Symbol used to mark luminaires	No symbol	⊡

Extracts from "BS4533, 1988" are reproduced with the permission of BSI under licence number 2000SK/0371. Complete British standards can be obtained by post from BSI Customer Services, 389 Chiswick High Road, London W4 4AL, United Kingdom (Tel. UK +020 8996 9001).

Luminaires are classified for the type of surface they can be mounted on. For example if a luminaire is only suitable to be mounted directly on non-combustible surfaces then a warning notice must be shown. If, however, a luminaire has a built-in ballast or transformer but is suitable for mounting on normally flammable surfaces then the symbol as shown in Figure 2.18 is used.

The BSI safety mark illustrated in Figure 2.16 is a guarantee of quality and confirms that the luminaire complies with the British Standard requirements.

*Figure 2.16   BSI safety mark
Reproduced with kind permission from BSI*

Temperature of certain objects or surfaces can be raised if the luminaire is too close to them.

Consequently a limit is put on the distance m between the luminaire and the surface which is lit (Figure 2.17).

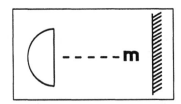

Figure 2.17   Minimum distance recommended between luminaire and lit surface (for example spotlights)

The F-Mark symbol shown on luminaires, indicates that they are suitable for direct mounting on normally flammable surfaces (Figure 2.18).

Figure 2.18   "F-Mark" symbol

Ingress protection (IP)

The IP system classifies the degree of protection given by an enclosure (luminaire). The protection is against the ingress of solid foreign bodies, including harmful ingress of liquids.

An IP classification consists of two numbers:
- the first number refers to the degree of protection against the ingress of solid foreign bodies (for example fingers, tools, wires and dust).
- the second number designates the degree of protection against harmful ingress of liquids.

The higher the two numerals, the greater the degree of protection the luminaire offers.

Table 2.3

First Number		Protection against the ingress of solid bodies
IP		Brief details
0		no protection
1		protected against solid objects greater than 50 mm diameter (for example accidental touch by hands)
2		protected against solid objects greater than 12 mm diameter and not exceeding 80 mm in length (for example fingers)
3		protected against solid objects greater than 2.5 mm diameter (for example tools and wires)
4		protected against solid objects greater than 1 mm diameter (for example small tools and small wires)
5		protected against dust (ingress of dust is not in sufficient quantity to interfere with satisfactory operation)
6		totally protected against dust (dust-tight)

Second Number		Protection against the ingress of liquids
IP		Brief details
0		no protection
1		protected against vertically falling drops of water (for example condensation)
2		protected against dripping water when the enclosure is tilted at any angle up to 15° from the vertical
3		protected against water sprays up to 60° from the vertical
4		protected against water sprayed from all directions
5		protected against low pressure jets of water from all directions
6		protected against strong jets of water or heavy seas (for example for use on ship decks)
7		protected against the effects of water immersion (ingress of water in a harmful quantity shall not be possible when the enclosure is immersed in water under defined conditions of pressure and time)
8		protected against continuous submersion in water under conditions which shall be specified by the manufacturer

For example, a RAINPROOF luminaire with protection from finger contact within the enclosure would be IP23, where number 2 signifies the protection against finger contact and the number 3 signifies protection against rain.

A luminaire which may be exposed to harmful dust and rain will be designated IP63, where number 6 signifies a dust-tight enclosure and number 3 signifies protection against the ingress of water sprays (for example rain).

If the type of protection is not classified, the omitted number is replaced by the letter X. For example a luminaire designated IP4X indicates (by the number 4) that the enclosure is protected against the ingress of solid objects greater than 1 mm diameter. Letter X indicates that there is no specific protection against the ingress of water.

Try this
1. What is meant by the abbreviation SHR?

2. State the quantity symbol for each of the following:
   (a) luminance

   (b) luminous intensity

3. The working plane inside a large office would normally be at _____ _____ .

4. Describe the symbol that indicates a Class 2 type luminaire.

Luminaires in hazardous areas

Code of Practice BS5345 (EN 60079) should be referred to for installations in hazardous areas (other than mining applications or explosive processing and manufacture).

Before a luminaire is selected for use in a hazardous area (for example the petrochemical industry) the following information is required:
1. the classification of the hazardous area (zone)
2. the temperature classification or T class – the maximum surface temperature marked on the luminaire must be less than the ignition temperature of the vapours or gases that may be present. (Table 2.5)

1. Classification of Hazardous Area

These areas are classified into three zones:

Zone 0   an area where explosive gas-air mixture is continuously present or present for long periods.

Zone 1   an area in which an explosive gas-air mixture is likely to occur in normal operation.

Zone 2   an area in which explosive gas-air mixture is not likely to occur in normal operation, and if it did occur, will only exist for a short time.

Temperature Classification

Table 2.5

T class	Maximum surface temperature °C
T1	450
T2	300
T3	200
T4	135
T5	100
T6	85

Luminaires certified by the British Approvals Service for Electrical Equipment in Flammable Atmospheres carry the BASEEFA mark shown in Figure 2.19.

The EEC mark for luminaires in hazardous areas are indicated by the distinctive Community Mark as shown in Figure 2.20.

Figure 2.19   BASEEFA mark

Selecting lamp and luminaire

Figure 2.20   Community mark

The choice of the lamp and luminaire combination for an indoor general lighting system should be based on the answers to the following questions:
1. What type of light distribution is required from the luminaire?
   (a) will the proportion of light on all surfaces be sufficient?
   (b) is there a danger of direct or indirect glare?
2. Are the colour rendering properties of the lamp suited for the intended use?
3. Is the lamp/luminaire reasonably efficient?
   (lamp efficacy and energy saving)
4. Does the luminaire harmonise with the decorations and surroundings?
5. Will the luminaire be suitable to withstand the physical conditions (for example moisture, dust, vibration, temperature, corrosion and vandalism)?
6. Is the luminaire suitable to operate in hazardous areas (zones) where explosive gases or liquids are present?
7. Will the lamp produce a stroboscopic effect if installed above moving machinery?
8. Will the luminaire be easy to clean and maintain?

Types of luminaires

Interior luminaires

There is such an enormous variety of luminaires made for interior applications that one can do no more than make general comments about them. However, they can be categorised into three types of applications, namely domestic, commercial and industrial luminaires.

Recessed luminaires
Swivel downlight luminaire

Typical applications

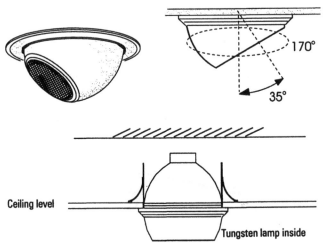

Figure 2.21

- Domestic and commercial, for example kitchen and lounge areas, shops, hotel receptions and exhibitions.

Surface mounted decorative luminaire tungsten lamps

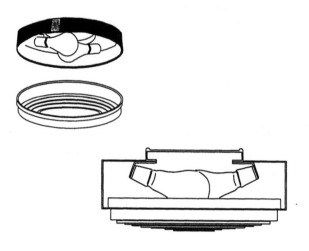

Figure 2.22

Typical applications

- Domestic and commercial, for example lounge areas, corridors, lobbies and hotels.

Track lighting

Extra low voltage tungsten halogen lamp luminaires supplied via a lighting track and transformer (Figure 2.23).

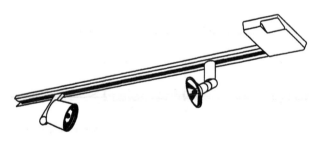

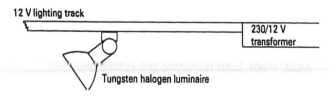

Figure 2.23

Typical applications

- Domestic and commercial, for example kitchens, shop window display, hotels, exhibitions and galleries.

Fluorescent luminaires

Fluorescent luminaires are manufactured to suit a range of interior applications. The control of the light from the lamp source will depend upon the lamp enclosure (for example prismatic controller). Controllers are available to suit the requirements of different types of installations. In hospitals, for example, centre ward lighting must be as glare-free as possible so that no discomfort is caused to the patients. Office lighting must have adequate levels of illuminance over the working plane (desk height) but again, glare must be avoided.

Surface mounted fluorescent luminaire

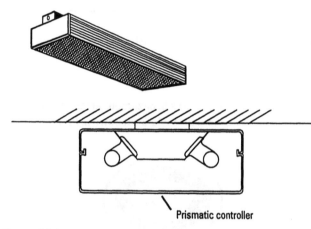

Figure 2.24

Typical applications

- A general purpose luminaire that is suitable for a wide range of domestic and commercial installations.

Fluorescent lighting trunking system

This is a versatile steel trunking system that is designed to support individual or continuous mounted fluorescent luminaires, as well as other electrical services. The trunking can be suspended or recessed (Figure 2.25).

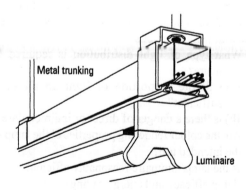

Figure 2.25

Typical applications

- Commercial and industrial, for example storage areas, factory interiors, exhibition halls.

High bay luminaire

This type of luminaire is available for use with high pressure sodium (SON), metal halide (MBI) or mercury vapour (MBF) lamps (Figure 2.26). Different light distributions are available to suit a wide range of visual requirements.

Figure 2.26

Typical applications

Commercial and industrial applications, for example large factory areas, transport terminals and sport halls.

Exterior luminaires

Luminaires are installed outdoors for many reasons.

For example;
- floodlighting of buildings and advertising boards
- motorway, road and street lighting
- sporting activities
- security of premises
- general area lighting (for example car parks, stock yards, docks and shopping precincts)

Luminaires selected for outdoor use must be weatherproof and where the need arises, vandal proof. The majority of outdoor luminaires installed for large area lighting schemes contain discharge lamps, mainly for their high lamp efficacy and low running cost as compared with tungsten filament lamps. However, tungsten halogen lamps are used for many outdoor applications and in situations where the colour rendering properties of discharge lamps would not be suitable.

Many factors that are taken into consideration during the design of an indoor lighting installation also apply to general lighting for outdoor schemes for example illuminance level, glare, maintenance and utilisation factors and spacing. For general area lighting however, further consideration must also be given to:

- siting of luminaires (on buildings or columns)
- adequate height of luminaires above traffic
- the effect of reflectance from road and building surfaces
- the control of luminaires (manual, time switch or photocell)
- the effect of shadows from any obstruction
- the ease of access for lamp replacement and maintenance

Discharge lamp luminaires

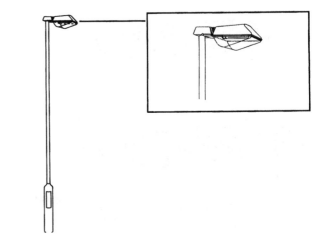

Figure 2.27

Typical applications

- suitable for major traffic routes, streets and car parks (Figure 2.27). The luminaire uses SOX, SON, MBF or MBI lamps.

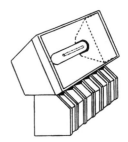

Figure 2.28   SON lamp floodlight

- floodlighting of buildings, stock yards, large car parks, security lighting and loading areas (wall or column mounted floodlights are also available) (Figure 2.28).

- used for amenity lighting such as gardens, courtyards, car parks, precincts and approach roads. The luminaire uses SON, SOX or MBF lamps (Figure 2.29).

Figure 2.29   Post top lantern

Fluorescent bulkhead luminaire

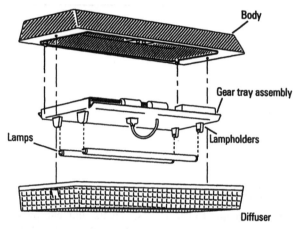

Figure 2.30

Typical applications

- The bulkhead luminaire is suitable for pedestrian walkways, security lighting and the like.

Tungsten halogen floodlight

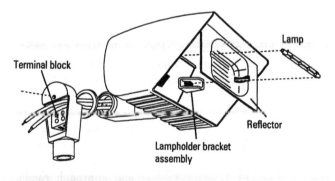

Figure 2.31

Typical applications

- The floodlight is suitable for a wide range of applications for example stock yards, car parks, building sites, forecourts and play areas.

Tungsten halogen floodlight and sensor

An exterior lighting installation that is designed to help protect property, plant and people from intruders may adopt the tungsten halogen floodlight combined with a detection sensor. Figure 2.32 illustrates this type of floodlight and passive infra-red motion detector (sensor). When a person or heat-radiating objects move within the detecting range of the sensor, it will automatically switch on the floodlight. It switches off again after a preselected time is reached. Adjustment is also available to vary the detecting range of the sensor and a twilight control is provided so that the floodlight will only operate during twilight or darkness.

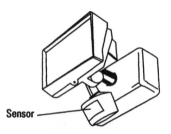

Figure 2.32

Typical applications

- many domestic, commercial and industrial uses, for example entrance and footpaths to domestic property, building sites, factory stock yards and entrance gates.

Hazardous area luminaires

The degree of hazard, mentioned earlier in the text, will dictate the choice of luminaire to be installed in the hazardous area (zone). Figures 2.33 and 2.34 show two different types of luminaires that are suitable for general lighting in hazardous areas.

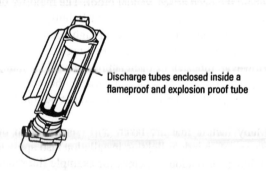

Figure 2.33   Zone 1 flameproof fluorescent luminaire

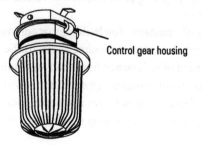

Figure 2.34   Wellglass luminaire for use in Zone 2 hazardous areas

Exercises

1. Polar curve data for a symmetrical luminaire is as follows:

Angle from the vertical plane in degrees	0°	5°	10°	15°	20°	25°	30°
Luminous intensity in candelas (cd)	400	440	450	420	280	80	0

   Construct the polar curve from the above data. State the luminous intensity at 17°.

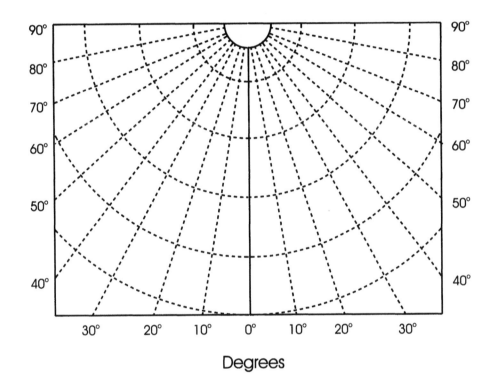

2. (a) State the purpose of using a utilisation factor when designing a lighting installation.

   (b) Describe the construction of a CLASS II luminaire.

3. The illumination of an exterior car parking area is in the process of being planned.
   (a) State four factors that should be considered during the planning stages.
   (b) State the discharge luminaire/lamp most suitable for the car park, giving the reason for your choice.

4. (a) Describe the effect on personnel of excess illuminance on the working plane in a drawing office.
   (b) Discuss in detail the benefits of group replacement for fluorescent lamps installed in a large clothes shop.

3

Laws of Illumination

Complete the following to remind yourself of some important facts on this subject that you should remember from previous work.

The diagram of a _____ curve will indicate the way _____ is distributed.

For an indoor lighting scheme the _____ of illuminance over an area is just one of the factors that must be considered during the _____ stages.

High bay luminaires can be used for _____ and _____ applications.

It is important that the _____ rendering of a lamp is suited for the area in which it operates.

When selecting a luminaire it must be _____ to withstand the _____ conditions in which it is installed.

Remember

LUMINOUS INTENSITY is measured in CANDELAS, symbol I

ILLUMINANCE is measured in LUX, symbol E

ONE LUX is also equal to ONE LUMEN per sq. metre

On completion of this chapter you should be able to:

- determine the illuminance at any point on a surface
- use polar curves with lighting calculations
- quote standard maintained illuminance for a selected sample of interiors
- determine the number of luminaires required for a given area
- produce scaled plans showing the correct spacing of luminaires

Figure 3.1

Inverse square law

The illuminance on a surface which is produced by a single light source, varies inversely as the square of the distance from the source. This is known as the INVERSE SQUARE LAW and is shown by the expression:

$$E = \frac{I}{d^2}$$

where:

- E = illuminance in LUX
- I = luminous intensity of the light source in CANDELA
- d = distance in metres from the light source to a point on a surface

The INVERSE SQUARE LAW is illustrated in Figures 3.2–3.4. An incandescent lamp of luminous intensity 500 candelas is fixed at different distances above a flat surface. The value of illuminance E on the surface is calculated for each distance d.

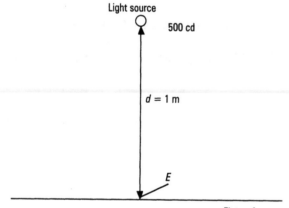

Figure 3.2

$$E = \frac{I}{d^2} = \frac{500}{1^2} = 500 \text{ lux}$$

Figure 3.3

$$E = \frac{I}{d^2} = \frac{500}{2^2} = 125 \text{ lux}$$

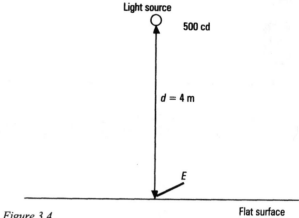

Figure 3.4

$$E = \frac{I}{d^2} = \frac{500}{4^2} = 31.25 \text{ lux}$$

If the distance is doubled between the light source and surface, the illuminance E will fall to one quarter of the previous value. Although the illuminated area will increase in size, the illuminance on the surface however will decrease accordingly. This effect can be demonstrated by shining a torchlight directly onto a flat surface and observing the increased area of light when the torch is moved further away from the surface.

Try this
An incandescent lamp with a luminous intensity of 350 cd in all directions is fixed directly above a table. The illuminance on the table top beneath the lamp is 75 lux.
(a) Determine the height of the lamp above the table.
(b) If the height of the lamp is increased by 50% determine the power of the source (cd) to produce the same illuminance.

Cosine law

If a beam of light from a lamp hits a surface at an angle, the illuminated area increases but the illuminance on the surface is lower than when the light is pointed directly at the surface. This effect can be demonstrated by holding a torchlight at an angle to a surface, rather than directly at the surface, and observing the increased illuminated area. The illuminance at a point on the surface will now be reduced by a factor of the cosine of the angle. This is known as the COSINE LAW and is shown by the expression:

$$E = \frac{I}{d^2} \times \cos\theta$$

The COSINE LAW is illustrated in Figure 3.5. A 500 cd incandescent lamp is fixed at a height of 2 metres directly above a long bench, and the value of illuminance at point P is to be determined.

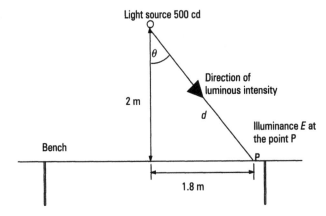

Figure 3.5

$$E = \frac{I}{d^2} \times \cos\theta$$

Distance d must be found:

$$d = \sqrt{2^2 + 1.8^2}$$

$$d = \sqrt{4 + 3.24}$$

$$d = \sqrt{7.24} \quad = 2.69 \text{ m}$$

Applying the Cosine Law:

$$E = \frac{500}{2.69^2} \times \frac{2}{2.69}$$

$$= 51.373 \text{ lux illuminance at P}$$

Try this

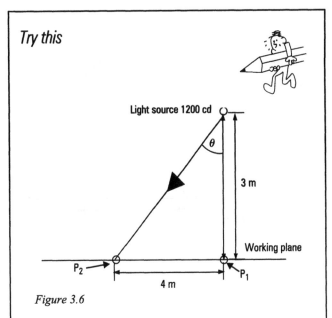

Figure 3.6

Determine the illuminance at points P_1 and P_2 on the working plane.

The previous examples illustrating the INVERSE SQUARE LAW and COSINE LAW have only referred to a single light source. Let us now consider various situations where there are several sources of light. For each situation the illuminance at a point is calculated from each source, and these values are added together to give the total illuminance at a point on a surface.

Example one

Figure 3.7 shows two luminaires, L_1 and L_2, mounted 5 metres apart. Each lamp inside the luminaires emits 600 cd in all directions. The values of illuminance at points P_1, P_2 and P_3 are to be determined.

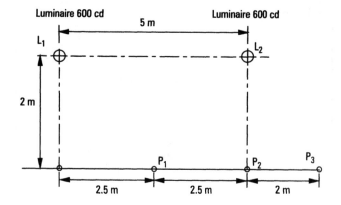

Figure 3.7

Illuminance at P₁

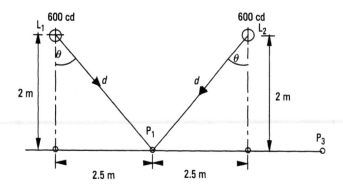

Figure 3.8

Distance d must be calculated:

$$d = \sqrt{2^2 + 2.5^2}$$
$$= \sqrt{4+6.25}$$
$$= \sqrt{10.25}$$
$$= 3.201 \text{ m}$$

Applying the Cosine Law to one lamp only:

$$E = \frac{I}{d^2} \times \cos\theta$$
$$= \frac{600}{3.201^2} \times \frac{2}{3.201}$$

$$= 36.586 \text{ lux}$$
illuminance due to one lamp only

Since the distance d to P_1 is the same from each luminaire, the total illuminance at P_1 is:

$$= 2 \times 36.586 \quad = 73.172 \text{ lux}$$

Illuminance at P₂

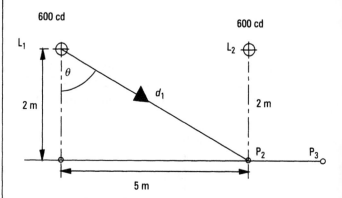

Figure 3.9

Applying the Inverse Square Law to lamp in L_2:

$$E = \frac{I}{d^2}$$
$$= \frac{600}{2^2}$$
$$= 150 \text{ lux}$$
illuminance due to L_2

Distance d_1 must be calculated:

$$d_1 = \sqrt{5^2 + 2^2}$$
$$= \sqrt{25+4}$$
$$= \sqrt{29}$$
$$= 5.385 \text{ m}$$

Applying the Cosine Law to lamp in L_1:

$$E = \frac{I}{d_1^2} \times \cos\theta$$
$$= \frac{600}{5.385^2} \times \frac{2}{5.385}$$

= 7.684 lux
illuminance due to L_1

Total illuminance at P_2 is:

= 150 + 7.684 lux
= 157.684 lux

Illuminance at P_3

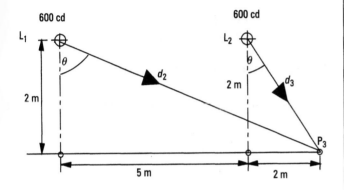

Figure 3.10

Distance d_2 must be calculated:

d_2 = $\sqrt{2^2 + 7^2}$

= $\sqrt{4 + 49}$

= 7.28 m

Applying the Cosine Law to lamp in L_1:

E = $\dfrac{I}{d_2^2} \times \cos\theta$

= $\dfrac{600}{7.28^2} \times \dfrac{2}{7.28}$

= 3.110 lux
illuminance due to L_1

Distance d_3 must be calculated:

d_3 = $\sqrt{2^2 + 2^2}$

= $\sqrt{4 + 4}$

= 2.828 m

Applying the Cosine Law to lamp in L_2:

E = $\dfrac{I}{d_3^2} \times \cos\theta$

= $\dfrac{600}{2.828^2} \times \dfrac{2}{2.828}$

= 53.057 lux

Total illuminance at P_3 is:

= 3.110 + 53.057 lux
= 56.167 lux

Try this
Determine the illuminance at points P_1 and P_2 on the working plane in Figure 3.11.

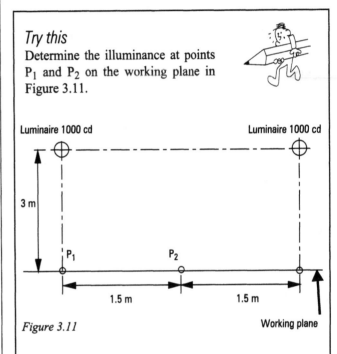

Figure 3.11

Example two

The forecourt of a building is to be illuminated by four floodlights (Figure 3.12) mounted on 6 metre high poles. A pole is positioned at each corner of the square forecourt which measures 20 m × 20 m. Each lamp and luminaire combination produces a luminous intensity of 2500 cd. Illuminance levels are to be determined at:

(a) the base of each pole, P_1
(b) the centre of the forecourt, P_2
(c) a point on the ground midway between each pole, P_3

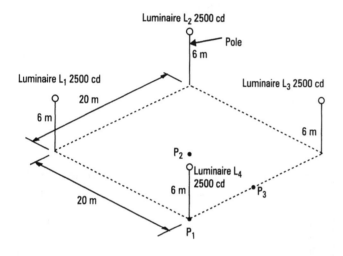

Figure 3.12

(a) Illuminance at P_1

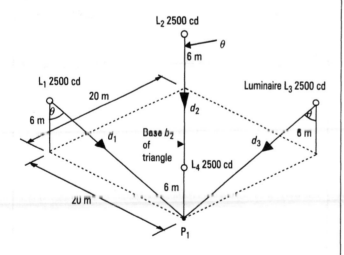

Figure 3.13

Distance d_1 must be calculated (Figure 3.13):

$$d_1 = \sqrt{6^2 + 20^2}$$

$$= \sqrt{36 + 400}$$

$$= \sqrt{436}$$

$$= 20.88 \text{ m}$$

Applying the Cosine Law to lamp in L_1:

$$E = \frac{I}{d_1^2} \times \cos\theta$$

$$= \frac{2500}{20.88^2} \times \frac{6}{20.88}$$

$$= 1.647 \text{ lux}$$
illuminance due to L_1

Illuminance due to L_3 will also be the same as L_1.

Therefore $L_1 + L_3 = 1.647 + 1.647$ lux
$= 3.294$ lux

To determine d_2, distance b_2 must be calculated first:

$$b_2 = \sqrt{20^2 + 20^2}$$

$$= \sqrt{400 + 400}$$

$$= \sqrt{800}$$

$$= 28.284 \text{ m}$$

$$d_2 = \sqrt{6^2 + 28.284^2}$$

$$= \sqrt{36 + 800}$$

$$= \sqrt{836}$$

$$= 28.913 \text{ m}$$

Applying the Cosine Law to L_2

$$E = \frac{I}{d_2^2} \times \cos\theta$$

$$= \frac{2500}{28.913^2} \times \frac{6}{28.913}$$

$$= 0.620 \text{ lux}$$
illuminance due to L_2

Applying the Inverse Square Law to L_4

$$E = \frac{I}{d^2}$$

$$= \frac{2500}{6^2}$$

= 69.444 lux

Total illuminance at P_1 is:

= 3.294 + 0.620 + 69.444 lux

= 73.358 lux

Since the area is a square, the illuminance at P_1 will be the same value at the base of all the poles.

(b) Illuminance at P_2

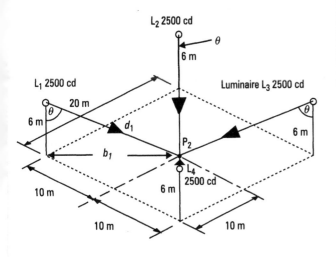

Figure 3.14

The illuminance at P_2 is the total illuminance produced by the four lamps. Since the spacings of all the lamps are the same, the total illuminance due to one lamp multiplied by four will provide us with the answer.

To determine distance d_1, distance b_1 must be calculated first (Figure 3.14):

b_1 = $\sqrt{10^2 + 10^2}$

= $\sqrt{100 + 100}$

= $\sqrt{200}$

= 14.142 m

Distance d_1

= $\sqrt{6^2 + 14.142^2}$

= $\sqrt{36 + 200}$

= $\sqrt{236}$

= 15.362 m

Applying the Cosine Law to L_1:

E = $\dfrac{I}{d_1^2} \times \cos\theta$

= $\dfrac{2500}{15.362^2} \times \dfrac{6}{15.362}$

= 4.137 lux

Total illuminance at P_2 is:

= 4.137 × 4

= 16.548 lux

(c) Illuminance at P_3

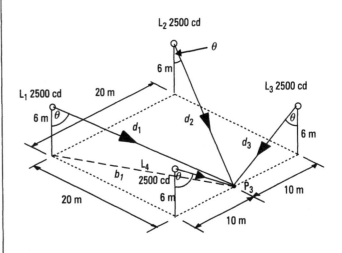

Figure 3.15

Illuminance produced by L_1 is the same as L_2.

To determine distance d_1, distance b_1 must be calculated first (Figure 3.15):

b_1 = $\sqrt{10^2 + 20^2}$

= $\sqrt{100 + 400}$

= $\sqrt{500}$

= 22.36 m

d_1 = $\sqrt{6^2 + 22.36^2}$

= $\sqrt{36 + 500}$

= $\sqrt{536}$

= 23.15 m

Applying the Cosine Law to L_1:

$$E = \frac{I}{d_1^2} \times \cos\theta$$

$$= \frac{2500}{23.15^2} \times \frac{6}{23.15}$$

$$= 1.209 \text{ lux}$$

Illuminance due to L_1 and L_2

$$= 1.209 \times 2$$
$$= 2.418 \text{ lux}$$

Illuminance produced by L_3 is the same as L_4.

distance
$$d_3 = \sqrt{6^2 + 10^2}$$
$$= \sqrt{36 + 100}$$
$$= \sqrt{136}$$
$$= 11.661 \text{ m}$$

Applying the Cosine Law to L_3:

$$E = \frac{I}{d_3^2} \times \cos\theta$$

$$= \frac{2500}{11.661^2} \times \frac{6}{11.661}$$

$$= 9.459 \text{ lux}$$

Illuminance due to L_3 and L_4

$$= 9.459 \times 2$$
$$= 18.918 \text{ lux}$$

Total illuminance at P_3 is:

$$= 2.418 + 18.918 \text{ lux}$$
$$= 21.336 \text{ lux}$$

Remember
The illuminance at a point on a surface is due to the **LUMINOUS FLUX** sent out by all the sources of light.

Try this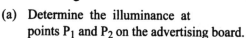

Advertising board

(a) Determine the illuminance at points P_1 and P_2 on the advertising board.
(b) Describe the type of luminaire you would select to illuminate the advertising board, giving the reasons for your choice.

The luminaire is situated on the centre line of the advertising board.

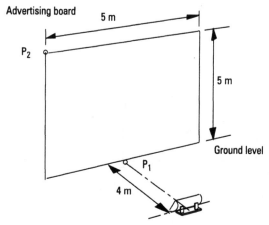

Figure 3.16   Luminaire fixed at ground level, giving a luminous intensity of 5000 cd in all directions.

Example three

The polar curve for a luminaire with an MBF lamp is shown in Figure 3.17. The luminaire is fixed at a height of 7 metres above a horizontal surface. Illuminance levels are required at:
(a) a point P_1 directly below the luminaire
(b) point P_2, 7 metres along the horizontal surface from P_1.

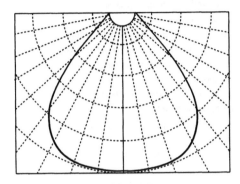

Figure 3.17   Polar curve

(a) Illuminance at P_1

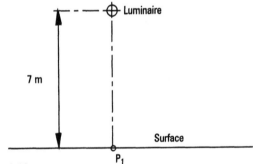

Figure 3.18

Applying the Inverse Square Law and the reading of 5000 cd at 0° on polar curve, illuminance at P_1 is:

$$E = \frac{I}{d^2}$$
$$= \frac{5000}{7^2}$$
$$= 102.04 \text{ lux}$$

(b) Illuminance at P_2

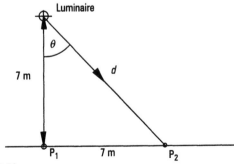

Figure 3.19

distance

$$d = \sqrt{7^2 + 7^2}$$
$$= \sqrt{49 + 49}$$
$$= \sqrt{98}$$
$$= 9.889 \text{ m}$$

$$\theta = 45°$$
$$\cos\theta = 0.707$$

From polar curve, the luminous intensity at 45° from the vertical is 3000 cd.

Applying the Cosine Law:

$$E = \frac{I}{d^2} \times \cos\theta$$
$$= \frac{3000}{9.899^2} \times 0.707$$
$$= 21.64 \text{ lux}$$
illuminance at P_2

Standard maintained illuminances for interiors

The selected data listed below provides guidance on the standard maintained illuminance required to suit various types of interiors. Illuminance values are based on considerations of the performance of appropriate activities, the general comfort of personnel involved in the activities, and the time the area is occupied.

	Standard Maintained Illuminance (LUX)
AGRICULTURE AND HORTICULTURE	
Inspection of Farm Produce where colour is important	500
Farm Workshops	
General	100
Workbench or machine	300
Milking parlours	100
Sick animal pens, calf nurseries	50
COMMERCE	
Offices	
General offices	500
Computer work stations	300–500
Conference rooms, executive offices	500
Computer and data preparation rooms	500
Filing rooms	300

Drawing offices
- General — 500
- Drawing boards — 750
- Print rooms — 300

Banks and Building Societies
- Counter, office area — 500
- Public area — 300

DISTRIBUTION AND STORAGE

Loading bays — 150

Work Stores
- Unpacking, sorting — 200
- Large item storage — 100
- Small item rack storage — 300
- Issue counter, records storeman's desks — 500

Warehouse and Bulk Stores
- Storage of goods where identification requires only limited perception of detail — 100
- Storage of goods where identification requires perception of detail — 150
- Automatic high bay rack stores:
  - gangway — 20
  - control station — 200
- Packing and despatch — 300

Cold Stores
- General — 300

EDUCATION

Assembly Halls
- General — 300

Teaching Spaces
- General — 300

Lecture Theatres
- General — 300
- Demonstration benches — 500

Seminar Rooms — 500
Art Rooms — 500
Needlework Rooms — 500
Laboratories — 500
Libraries — 300
Sports Halls — 300
Workshops — 300

HOSPITALS AND HEALTH CARE BUILDINGS

Corridors
- General — 150
- Ward Corridors open to daylight:
  - day — 150–200
  - night — 3–5

Wards
- Circulation
  - day — 100
  - night — 3–5
- Operating and Treatment Areas
  - Operating Room:
    - general — 400–500
    - cavity — 10000–50000

Endoscopy — 400
Scrub-up Room — 300
Recovery Room — 300

MECHANICAL ENGINEERING

Structural Steel Fabrication:
- General — 300
- Marking off — 500

Sheet Metal Works
- Pressing, punching, shearing, stamping, spinning, folding — 500
- Benchwork, scribing, inspection — 750

Welding and Soldering Shops:
- Gas and arc welding, rough spot welding — 300
- Medium soldering, brazing, spot welding — 500
- Fine soldering, fine spot welding — 1000

Assembly Shops:
- Rough work, e.g. frame and heavy machinery assembly — 300
- Medium work, e.g. Engine assembly, vehicle body assembly — 500
- Fine work, e.g. office machinery assembly — 750
- Very fine work, e.g. instrument assembly — 1000
- Minute work, e.g. watch making — 1500

Paint Shops and Spray Booths:
- Dipping, rough spraying — 300
- Preparation, ordinary painting, spraying and finishing — 500
- Fine painting, spraying and finishing — 750
- Inspection, retouching, matching — 1000

PLACES OF PUBLIC ASSEMBLY

Assembly Rooms
- Public rooms, village halls, church halls — 300

Concert halls, Cinemas and Theatres
- Foyers — 200
- Booking office — 300
- Auditoria — 100–150
- Dressing rooms — 300
- Projection room — 150

Places of Worship
- Body of church — 100–200
- Pulpit, lectern — 300
- Choir stalls — 200
- Altar, communion table, etc. — 300
- Vestries — 150
- Organ — 300

RESIDENTIAL BUILDINGS

Hotels (Public and Staff Areas)
- Entrance halls — 100
- Reception, cashiers' and porters' desks — 300
- Bars, coffee bars, dining rooms, grill rooms, restaurants, lounges — 50–200
- Cloakrooms, baggage rooms — 100
- Guest bedrooms/lounges — 50–100
- Guest Bathrooms — 150

TIMBER AND FURNITURE

Sawmills
- General — 200
- Head saw — 500
- Grading — 750

Woodwork Shops
- Rough sawing, benchwork — 300

Sizing, planing, sanding, medium machining and benchwork	500
Fine bench and machine work, fine sanding, finishing	750
Furniture Manufacture	
Raw materials stores	100
Finished goods stores	150
Wood matching and assembly, rough sawing, cutting	300
Machining, sanding and assembly, polishing	500
Tool rooms	500
Spray booths:	
colour finishing	500
clear finishing	300
Cabinet making:	
veneer sorting and grading	1000
marquetry, pressing, patching and fitting	500
Final inspection	750
Upholstery Manufacture	
Cloth inspection	1500
Filling, covering	500
Slipping, cutting, sewing	750
Mattress making:	
assembly	500
tape edging	1000

For a more detailed list of lighting recommendations, reference should be made to the Chartered Institution of Building Service Engineers "Code for Interior Lighting".

The above selected data is reproduced from CIBSE *Code for Interior Lighting*, by permission of the Chartered Institution of Building Services Engineers.

Remember
The type of interior (environment) will govern the choice of luminaires.

Lighting design

The aim of this part of the workbook is to provide a general outline of interior lighting design. It must be emphasised that the information given is only sufficient to demonstrate the basic principles of lighting design calculations. Nevertheless, the information may be suitable for the planning of simple interior lighting schemes.

An increasing use is being made of computer packages for lighting schemes. Once the required data has been fed into the computer, accurate information of the number of luminaires required and their general layout is quickly printed out.

Remember
UTILISATION FACTOR is the proportion of luminous flux which is emitted by the lamps that reaches the working plane, taking into consideration the room reflectances, room dimensions and the height of luminaires above the working plane.

MAINTENANCE FACTOR allows for the depreciation of illuminance over a period of time.

STANDARD MAINTAINED ILLUMINANCE is the recommended level of illuminance in LUX for the assumed standard conditions of the application.

SPACING TO HEIGHT RATIO is the spacing between the centres of luminaires divided by their height above the working plane

In addition to the lighting terms discussed in Chapter 2 of this book, you will need to know the following when planning a lighting scheme:

ROOM INDEX (RI)

is related to the room dimensions and used when calculating the utilisation factor and other characteristics of a lighting installation.

$$RI = \frac{L \times W}{Hm(L+W)}$$

where: L = Length of room
W = Width of room
Hm = Mounting height of luminaire above the working plane

REFLECTION FACTORS OF ROOM SURFACES

take into consideration the reflection of illuminance from ceilings, walls and floor.

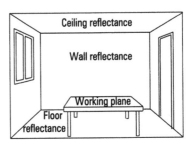

Figure 3.20

Very generalised reflection factors are:

70%	White finish or near white
50%	Light colour finish
30%	Medium colour finish
10%	Dark colour finish

Lumen method of calculation

The level of illuminance in places such as industrial workrooms and offices is usually prescribed in terms of the average illuminance on a horizontal working plane. To design a lighting scheme that will produce the desired level of illuminance, the quantity of luminaires must be determined. This is performed by a calculation known as the lumen method.

The formula is:

$$N = \frac{E_{av} \times L \times W}{F \times UF \times MF}$$

where

- N = Number of luminaires required
- F = The initial lumens of each lamp obtained from manufacturers' data multiplied by the number of lamps in each luminaire
- UF = Utilisation factor
- MF = Maintenance factor
- L = Length of room
- W = Width of room
- E_{av} = Average illuminance required in LUX (see recommended values of standard maintained illuminance, pages 39–41)

In order to find the utilisation factor, the room index must be calculated first. Having done this, assessment is made of the room reflectances. Both room index and reflectance factors are then applied to the manufacturers' photometric data to determine the utilisation factor for the luminaire.

For example:

Calculated Room Index = 2.41
(read 2.5 on data table)

Room reflectances C = 0.70, W = 0.30, F = 0.20

From photometric data table, utilisation factor for the luminaire is 0.46 (Table 3.1).

Table 3.1

Photometric data
UTILISATION FACTORS

One lamp diffuser

Room reflectance			Room index						
C	W	F	0.75	1.00	1.25	1.50	2.00	2.50	3.00
.70	.50	.20	–	.35	.40	.43	.48	.51	.53
	.30		–	.30	.34	.38	.43	.46	.49
	.10		–	.26	.30	.34	.39	.43	.46
.50	.50	.20	–	.32	.35	.38	.42	.45	.47
	.30		–	.27	.31	.34	.39	.42	.44
	.10		–	.24	.28	.31	.35	.39	.41
.30	.50	.20	–	.28	.31	.34	.37	.40	
	.30		–	.25	.28	.31	.34	.37	
	.10		–	.22	.25	.27	.32		
.00	.00	.00	–	.18	.21	.23	.26		

For the lamp wattage used, multiply above UF values by the factors given here:

18 W	20 W	36 W	40 W	58 W	65 W	70 W	75 W	100 W	125 W
–	–	1.06	1.04	1.03	1.0	1.03	1.0	0.99	0.98

If a calculated value of Room Index falls in between the values given in the photometric data table, an average can be taken between the upper and lower values of RI.

For example:

Calculated RI of 2.25 can be assumed to have a UTILISATION FACTOR halfway between RI 2.0 and 2.5.

In most cases however, after the calculated value of RI is found, discretion is used when selecting the appropriate Room Index from the photometric data table.

The following examples describe the procedure to adopt when planning a lighting scheme.

Example four

An office area requires an average illuminance of 500 lux on the working plane, 0.75 metres from the floor. Twin 75 watt fluorescent luminaires, 1800 mm long with an opal diffuser are to be fixed directly to the ceiling. Each lamp has a rating of 5600 lumens. Office dimensions are 12 m long by 6 m wide. Ceiling height is 2.68 metres and painted white. Walls have also a light finished surface.

Room surface reflection factors are:

Ceiling C = 0.70
Walls W = 0.50
Floor F = 0.20

Assume a maintenance factor to be 0.85.

The total number of luminaires for the office must be determined, together with a plan view of their spacing arrangement.

Table 3.2

Photometric data
UTILISATION FACTORS

Two lamp opal diffuser

Room reflectance			Room index						
C	W	F	0.75	1.00	1.25	1.50	2.00	2.50	3.00
.70	.50	.20	.32	.36	.41	.44	.48	.52	.54
	.30		.26	.31	.36	.39	.44	.48	.50
	.10		.23	.27	.32	.36	.41	.44	.47
.50	.50	.20	.28	.32	.36	.39	.43	.45	.47
	.30		.24	.28	.32	.35	.39	.42	.44
	.10		.21	.25	.29	.32	.36	.40	.42
.30	.50	.20	.25	.28	.32	.34	.37	.39	
	.30		.21	.26	.28	.31	.35	.37	
	.10		.19	.22	.26	.28	.32		
.00	.00	.00	.15	.18	.21	.23	.26		

For the lamp wattage used, multiply above UF values by the factors given here:

18 W	20 W	36 W	40 W	58 W	65 W	70 W	75 W	100 W	125 W
-	-	1.06	1.04	1.03	1.0	1.03	1.0	0.99	0.98

Table 3.3 Photometric Data

Type	Lamp watts	SHR (Nom.)	SHR (Max.)
Opal diffuser	1 × 65	1.75	1.95
	2 × 65	1.5	1.72
Prismatic controller	1 × 65	1.75	1.77
	2 × 65	1.5	1.73

Procedure

Step (1) Determine height of luminaire above the working plane.
Height of luminaire = 2.68 − 0.75 = 1.93 m

Step (2) Determine Room Index

$$RI = \frac{\text{Room length} \times \text{Room width}}{\text{Luminaire height above working plane} \times (\text{Room length} + \text{Room width})}$$

$$RI = \frac{12 \times 6}{1.93 \times (12 + 6)}$$

$$= 2.072 \ (2.0 \text{ approx.})$$

Step (3) Determine the Utilisation Factor for the luminaire.

From the photometric data table (Table 3.2) for this luminaire, read off Utilisation Factor, where C = 0.70, W = 0.50 F = 0.20 and Room Index = 2.0

Therefore, Utilisation Factor = 0.48

Step (4) Using the lumen method calculation, determine the number of luminaires required.

$$N = \frac{E_{av} \times L \times W}{F \times UF \times MF}$$

$$= \frac{500 \times 12 \times 6}{(5600 \times 2) \times 0.48 \times 0.85}$$

$$= 7.878 \text{ luminaires}$$

Obviously 7.878 is impossible, so 8 luminaires will be required for a symmetrical layout.

Step (5) Determine the Spacing of Luminaires

From photometric data, space-height ratio SHR (MAX) must not be exceeded (1.72 for a twin lamp luminaire with opal diffuser) (Table 3.3).

$$\frac{\text{Space between each luminaire S}}{\text{Height of luminaires above working plane H}} = 1.72 \text{ max.}$$

Maximum spacing between centres of each luminaire S = 1.93 × 1.72

= 3.319 metres

Step (6)

Draw the symmetrical layout for 8 luminaires so that the spacing S between centres of each luminaire does not exceed 3.319 m. The spacing between side or end luminaires and the wall should not be more than 0.5 × S (measuring from centre of luminaire to wall).

The layout of the luminaires can be drawn to a chosen scale. Spacing between centres of each luminaire should not exceed the maximum spacing allowed, otherwise unacceptably uneven levels of illuminance could occur (Figure 3.21).

The calculations used for the spacings of the fluorescent luminaires in this example are in conjunction with SPACE-HEIGHT RATIO data shown by the lighting manufacturer on Table 3.3. However, other manufacturers may show their data for the spacing of linear fluorescent luminaires in a different form, for example SHR MAX and SHR MAX TR. Whichever form the space-height ratio is shown by the manufacturer, their values must be complied with.

Notes

1. Apply any correction factors given in the photometric data to the UF value. However, in Example 4 the correction factor is 1.0, so that the UF remains the same.
2. Three space-height ratios may be quoted in the manufacturers' photometric data:
   SHR NOM (a nominal standard by which the utilisation factors in the tables are calculated), SHR MAX(IMUM) and SHR MAX(IMUM) TR(ANSVERSE).
   The MAXIMUM values of the SHR in the photometric data must not be exceeded. It is also important to realise that the SHR data is only calculated for unobstructed areas. Allowance may have to be made where screens or tall cabinets are placed in rooms (increase the number of luminaires and reduce spacing distances).
3. If the number of luminaires (chosen luminaires/lamps) calculated suggests spacings greater than that recommended (distance between luminaire centres), it will be necessary to recalculate, using more luminaires/lamps of slightly lower lumen output. Therefore, the uniformity and recommended illuminance for the area is maintained.
4. Glare must be avoided when planning a lighting scheme. However, the calculation of glare index is beyond the scope of this studybook.

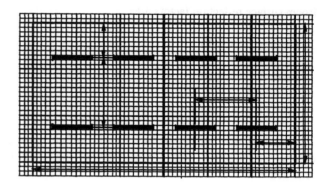

Figure 3.21   Plan view of luminaires

Try this

A school classroom is to be illuminated with single-lamp fluorescent luminaires fixed directly to the ceiling. Classroom dimensions measure, 12 metres long, 8 metres wide and 3 metres high. Desk height, 0.7 metres.

Photometric data tables show that:

UTILISATION FACTOR	= 0.66
INITIAL LUMENS of each lamp	= 4600 lm
SHR MAXIMUM	= 1.9

(a) (i) State the recommended average illuminance required for the classroom (teaching spaces).
   (ii) Use the value in (i) to determine the number of luminaires required. Assume a maintenance factor of 0.86.
(b) Determine the maximum spacing distance between centres of each luminaire.

Example five

A small factory requires an average illuminance of 300 lux at bench level, 0.87 metres from the floor. Factory dimensions are 20 metres long, 10 metres wide and 8 metres high. SON 250 watt discharge lamps in high bay luminaires are to be used. The luminaires are mounted 1 metre from the ceiling. Assume a maintenance factor of 0.61.

Photometric data tables show that:

UTILISATION FACTOR	= 0.57
SHR MAXIMUM	= 1.13
INITIAL LUMENS OF EACH LAMP	= 24000 lm

The total number of luminaires must be determined, and their spacing arrangement.

$$N = \frac{E_{av} \times L \times W}{F \times UF \times MF}$$

$$= \frac{300 \times 20 \times 10}{24000 \times 0.57 \times 0.61}$$

= 7.19 luminaires
Actual number will be 8 luminaires.

Determine height of luminaires above the bench:

= Ceiling height − luminaire distance from ceiling − bench height

= 8 − 1 − 0.87
= 6.13 metres

Determine the spacing of luminaires

$$\frac{S}{H} = 1.13 \text{ maximum}$$

$$\frac{S}{6.13} = 1.13$$

Maximum spacing between centres of luminaires

S = 6.13 × 1.13

= 6.926 metres

Two rows of luminaires with four luminaires in each row.

Obviously, the spacing between each luminaire will not exceed the maximum value.

Try this

Determine the average illuminance in a room 8 metres by 5 metres, if six twin-lamp fluorescent luminaires are installed.

Photometric data tables show that:

UTILISATION FACTOR	= 0.47
INITIAL LUMENS of each lamp	= 4200 lm

Assume a maintenance factor of 0.8.

Try this

Draw a plan view of the spacing arrangement for the high bay luminaires in Example five.
Use a scale of 10 mm = 1 metre
Indicate the centre of each luminaire by means of a +.

4

Lighting Loads and Special Systems

Complete the following to remind yourself of some important facts on this subject that you should remember from the previous chapter:

The two laws that can be used to find the illuminance at a point are called _____ and _____ .

Complete the following formula

$$E = \frac{I}{d^2} \times \underline{\qquad}$$

The abbreviation RI represents

_____ _____

A luminaire controls the light distribution of a lamp and this can be illustrated by means of a _____ _____ diagram.

The approximate reflection factor from a white surface wall is _____ .

On completion of this chapter you should be able to:

- state the causes and effects of harmonics
- determine the kW, kVA_r, kVA, power factor and current of a lighting load
- determine the maximum length of run for a given cable size connected to extra low voltage lamps
- identify the Regulations (BS7671) associated with high voltage discharge sign installations
- state the requirements for emergency lighting installations and the type of systems available
- describe air handling luminaires

Harmonics

The expression, HARMONIC MOTION, can be applied to all types of periodic oscillatory motion. For example, the vibration of the strings from musical instruments (from which association the term HARMONIC is derived), the movement of a pendulum, and alternating, or oscillating, electrical or electromagnetic waves.

Electrical waveforms of voltage or current are often used to show how their values vary with time. The simplest and ideal wave is called a true SINE WAVE, as shown in Figure 4.1.

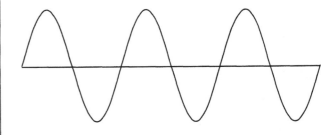

Figure 4.1   A single sine wave

However, a wave form which may approximate to a SINE WAVE, without being a true one, is said to be a distorted wave. The distortion of the sine wave is due to other sine waves being superimposed on it which have different amplitudes and periods. Those waveforms which produce distortion are called HARMONICS. A distorted waveform which is made up of a number of harmonics is called a complex wave. Harmonics have a frequency which are an exact multiple of the fundamental (or lowest) frequency. For example, the second harmonic has twice the frequency of the fundamental, as shown in Figure 4.2.

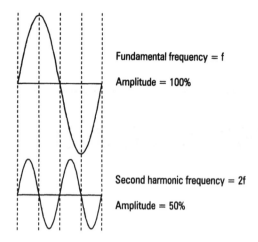

Figure 4.2

The third harmonic has three times the frequency of the fundamental, and so on. Figure 4.3 shows the fundamental frequency and the third harmonic.

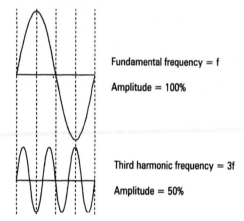

Figure 4.3

Complex waveforms

A complex waveform is the sum of the fundamental and harmonic frequency (or frequencies). Figure 4.4 shows the result of adding the SECOND HARMONIC to the fundamental. It can be seen that the resultant complex waveform has dissimilar positive and negative half-cycles.

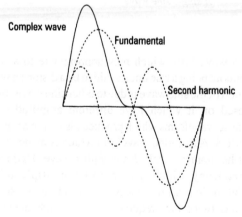

Figure 4.4

The resultant complex waveform produced by adding the THIRD HARMONIC to the fundamental, is shown in Figure 4.5. In this case, the complex waveform has identical positive and negative half-cycles.

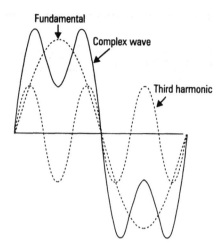

Figure 4.5

If a complex waveform contains the fundamental and an odd harmonic (third, fifth and so on), this always gives rise to symmetrical resultant waves in which positive and negative halves are identical. Even harmonics however, give rise to unsymmetrical waves.

Cause and effect of harmonics

Although it is desirable to have sinusoidal waveforms of currents and voltages in an a.c. distribution system, it may not always be possible, and harmonics injected into the system from one source or another will produce complex waveforms. Sources which may introduce harmonics into the supply system include discharge lighting circuits, electronic equipment and solid state controls (for example dimmer switches and motor speed controllers), alternators, transformers and the like.

Inductors and capacitors that are subjected to complex waveforms of this nature will have a wide variation of reactance at one of the harmonic frequencies. The possibility of resonance cannot be ruled out, which may result in high values of harmonic voltages across the inductor and capacitor, including a large harmonic circuit current.

Harmonic currents which are present in the mains supply to discharge lighting may be further increased by the power factor correction capacitors, due to the fact that they will have a lower impedance to the higher frequency harmonics in the waveforms.

Where discharge lighting circuits are connected to three phase 4-wire supply systems and a third harmonic current is present in each of the phases, the currents do not cancel out at the star point of the supply, since they are all in phase, as shown in Figure 4.6.

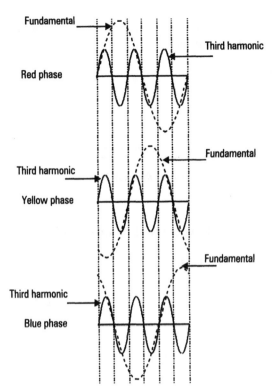

Figure 4.6   Presence of the third harmonic in each place of a three phase 4-wire supply system

In this situation the neutral conductor will have to carry three times the third harmonic current. So even on balanced loads the value of harmonic current in the neutral conductor could be quite high.

Where the neutral conductor of a polyphase circuit is likely to carry this harmonic current, its cross-sectional area shall be of adequate size (see 524-02 BS7671). For example, the cross sectional area of the neutral conductor feeding a 3-phase and neutral distribution board that is supplying fluorescent luminaires, must not be reduced below that of the phase conductors.

Equipment and controls which introduce harmonics that distort the fundamental sine waveform of the supply can be detrimental to other electrical equipment or services. Chapter 33 "COMPATIBILITY" of BS7671 points out that an assessment should be made of any characteristics of electrical equipment which are likely to have some form of harmful effect upon other equipment, or other services or supply.

Try this
1. (a) Explain the term: complex waveform

   (b) Name two types of electrical waveform.

2. State two disadvantages of harmonics in a.c. supply systems.

Lighting loads

Inductive loads

One of the disadvantages of discharge lamp circuits controlled by a choke or transformer, is the inductive current taken by these components. Although discharge luminaires have power factor correction capacitors connected to reduce most of the inductive current, circuit conductors will still be required to carry the remainder of this current. Figure 4.7 shows the power factor correction capacitor connected in a fluorescent lighting circuit.

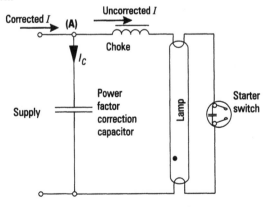

Figure 4.7   Inductive load and power factor correction

The leading current I_C taken by the capacitor will reduce (or correct) the inductive current in the circuit conductors to the left of position (A). Conductors to the right of position (A) will still carry the inductive current (uncorrected I). Hence, a capacitor will only correct the inductive current in the cables and switchgear in the direction of the mains supply position of the premises. The value of capacitance will determine the amount of correction that takes place in the current carrying conductors and switchgear.

Industrial and commercial premises which have a large amount of discharge lighting can reduce their total current, kVA demand and running costs by maintaining a high level of power factor for this type of load.

Circuit current

Circuits supplying discharge lighting must be capable of carrying the total steady current of the lamp and associated control gear, including their harmonic currents. Where more exact information is not available and the resulting power factor of the circuit is not less that 0.85 lagging, the current demand can be determined from the following formula:

$$\text{Current (I)} = \frac{\text{Lamp power (Watts)} \times 1.8}{\text{Supply voltage}} = \frac{\text{volt - amperes}}{\text{voltage}}$$

The 1.8 multiplier applied to the lamp watts is the projected demand in volt-amperes.
(Reproduced from the IEE On-Site Guide with kind permission of the IEE.)

For example:
A circuit supplying 12 × 75 watt fluorescent luminaires connected to a 230 V supply would take a current of:

$$I = \frac{12 \times 75 \times 1.8}{230} = 7.04 \text{ A}$$

Note:
For convenience "pf" is used as an abbreviation for power factor throughout this section.

Example one
The following lighting loads are balanced across a three phase, 4-wire 400/230 V supply:

Load 1   10 kW of incandescent lamps

Load 2   7 kW of MBF lamps at 0.89 pf lagging

Load 3   60 × 65 watt fluorescent lamps at 0.86 pf lagging

Determine the
(a) total kW
(b) total kVA$_r$
(c) total kVA
(d) overall pf
(e) line current

To determine the total values, the kW, kVA$_r$ and kVA for each load must first be calculated.

kW of load 1

$$kW = 10$$

Figure 4.8   Power phasor for load 1

NOTE
Phasor diagrams shown are NOT drawn to scale.

kVA and kVA$_r$ of load 2

$$kVA = \frac{kW}{pf} = \frac{7}{0.89} = 7.865 \text{ kVA}$$

$$kVA_r = \sqrt{kVA^2 - kW^2}$$

$$= \sqrt{7.865^2 - 7^2}$$

$$= \sqrt{61.858 - 49}$$

$$= \sqrt{12.858}$$

$$= 3.585 \text{ kVA}_r$$

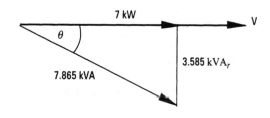

Figure 4.9   Power phasor for load 2

kW, kVA and kVA$_r$ of load 3

$$kW = \frac{60 \times 65}{1000} = 3.9 \text{ kW}$$

$$kVA = \frac{kW}{pf} = \frac{3.9}{0.86} = 4.534 \text{ kVA}$$

$$kVA_r = \sqrt{kVA^2 - kW^2}$$

$$= \sqrt{4.534^2 - 3.9^2}$$

$$= \sqrt{20.557 - 15.21} = \sqrt{5.347}$$

$$= 2.312 \text{ kVA}_r$$

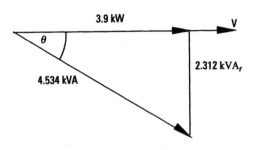

Figure 4.10   Power phasor for load 3

Determine the totals

Now resolve the total active (in-phase) and reactive (out of phase) components of the lighting load.

LOAD	kW (active)	kVA$_r$ (reactive)
i	10	–
ii	7	3.585
iii	3.9	2.312
Total	20.9	5.897

(a) Total kW = 20.9
(b) Total kVA$_r$ = 5.897

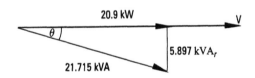

Figure 4.11   Combined lighting load phasor

(c) Total kVA = $\sqrt{kW^2 + kVA_r^2}$

$\qquad = \sqrt{20.9^2 + 5.897^2}$

$\qquad = \sqrt{436.81 + 34.774}$

$\qquad = \sqrt{471.584}$

$\qquad = 21.715$ kVA

(d) Overall pf = $\dfrac{\text{Total kW}}{\text{Total kVA}}$

$\qquad = \dfrac{20.9}{21.715} =$ 0.962 pf lagging

(e) Line current = $\dfrac{\text{kVA} \times 1000}{V_L \times \sqrt{3}}$

$\qquad = \dfrac{21.715 \times 1000}{400 \times 1.732}$

$\qquad = 31.34$ A

Try this
Find the value of line current taken by EACH of the following lighting loads, balanced across a three phase 4-wire 400/230 V supply:

Load 1   10 kW at 0.8 pf lagging
Load 2   15 kW of fluorescent luminaires
Load 3   8 kVA of low pressure sodium luminaires

Example two

An office measuring 20 metres by 15 metres requires an average illuminance of 500 LUX. 70 W fluorescent lamps, giving 92 lumens/watt are to be used.

(a) Determine the number of lamps required, assuming the utilisation and maintenance factors are 0.58 and 0.86 respectively.

(b) Calculate the current taken by each core of a three phase and neutral cable supplying the lighting fuseboard, assuming the load is balanced across the 400/230 V supply.

Answer

(a) Total lumens required from all lamps

$$= \frac{\text{Lux} \times \text{Area}}{\text{U.F.} \times \text{M.F.}}$$

$$= \frac{500 \times 20 \times 15}{0.58 \times 0.86}$$

$$= 300721.73 \text{ lumens}$$

Total power taken

$$= \frac{\text{Total lumens}}{\text{Lamp lumens per watt}}$$

$$= \frac{300721.73}{92}$$

$$= 3268.71 \text{ W}$$

Number of lamps required

$$= \frac{\text{Total power taken}}{\text{Lamp watts}}$$

$$= \frac{3268.71}{70}$$

$$= 46.69 \text{ (say 47 lamps)}$$

In practice 48 lamps would be used.

(b) Current taken by each core of the cable

$$= \frac{\text{Total lamp watts} \times 1.8}{\text{Phase voltage} \times 3}$$

$$= \frac{48 \times 70 \times 1.8}{230 \times 3}$$

$$= 8.765 \text{A}$$

Try this

A storage area requires a total of 400 000 lumens from 150 W high pressure sodium lamps. Lamp efficacy is 100 lumens per watt and the overall power factor is 0.9 lagging. The supply to the premises is three phase 4-wire 400/230 V.

(a) Determine the average level of illuminance over the area which measures 50 metres by 20 metres. Utilisation and maintenance factors are 0.56 and 0.8 respectively.

(b) Calculate the number of high pressure sodium lamps required.

(c) Determine the line current taken by the sub-main cable supplying the lighting distribution fuseboard.

(Note: The small power loss in the control gear has been ignored.)

Solution of lighting loads by phasor diagram to scale.

The total lighting loading of a building interior can also be determined by using an accurate scaled phasor diagram. Great care should be taken in the measurement of each quantity (use a fine tip pencil). A scaled phasor diagram can also prove most useful when checking calculated solutions (Figure 4.12).

Example three

The following lighting loads are balanced across a three phase, 4-wire 400/230 V supply:

Load 1   8 kW at unity power factor (incandescent lamps)
Load 2   7 kVA at 0.6 pf lagging (sodium lamps)
Load 3   5 kVA at 0.8 pf lagging (fluorescent lamps)

Determine by a scaled phasor diagram:
(a) total kW
(b) total kVA$_r$
(c) total kVA
(d) overall pf
(e) line current (this must be calculated).

SCALE: 1 cm = 1 kW
         1 cm = 1 kVA$_r$
         1 cm = 1 kVA

Solution to example three.

Determining phase angles required for phasor diagram (Figure 4.12):

$\cos \theta_1 = 0.8$
$\theta_1 = 36.86°$

$\cos \theta_2 = 0.6$
$\theta_2 = 53.13°$

Answers from phasor diagram:
(a) OA = Total kW = 16
(b) AB = Total kVA$_r$ = 8.8
(c) OB = Total kVA = 18.4
(d) Overall pf
    θ_3 = 28°
    $\cos 28°$ = 0.882 lagging

By calculation:

(e) $I_L = \dfrac{kVA \times 1000}{V_L \times \sqrt{3}}$

$= \dfrac{18.4 \times 1000}{400 \times 1.732}$

$= 26.558$ A

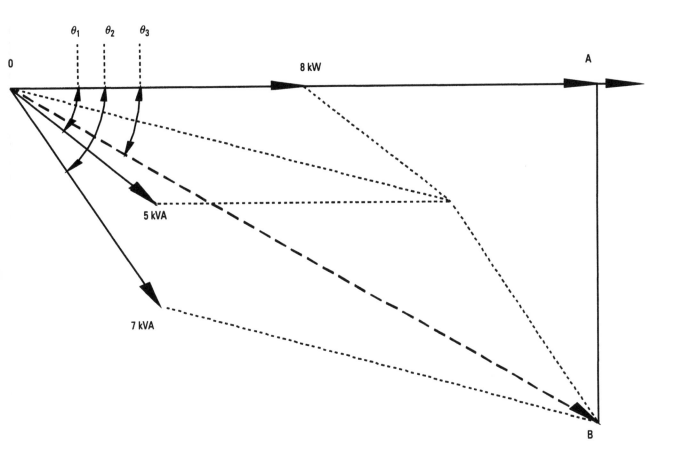

Figure 4.12   Solution by phasor diagram

Try this

The lighting load of a large factory consists of a mixture of incandescent lamps and discharge lamps. The load, which is balanced across the three phase, 4-wire 400/230 V supply, comprises:

Load 1   20 kW of discharge lamps at 0.8 pf lagging

Load 2   30 kW of tungsten halogen lamps

Determine, by a phasor diagram drawn to scale, the:
(a) total kW
(b) total kVA
(c) overall phase angle and power factor
(d) kVA_r rating per phase of a capacitor to reduce the phase angle to 5° lagging.

Balancing loads

Where three phase, 4-wire 400/230 V supply systems are brought into a building it is common practice to balance all the single-phase loads as evenly as possible across the different phases. Single-phase lighting circuits in particular, are often one of the largest sustained loads of commercial and industrial premises that should be distributed across the three phases. Careful planning of the phasing and final circuit arrangement in the different areas of the premises will ensure that, in the event of one phase or circuit failing, little disruption as possible is caused within the premises.

For example, the distribution system for an office lighting load in Figure 4.13 shows the arrangement of three single-phase lighting fuseboards on each floor. The supply to each fuseboard is taken from a 3-phase and neutral main distribution board. If one phase fails, there will still be lighting available in the majority of areas on each floor.

Each single-phase lighting fuseboard should be located as near as possible to the centre of the lighting load area it serves. By doing so, final circuit cable runs will be kept as short as possible and any drop in voltage is kept to a minimum. Compliance with BS7671, Section 525 (Voltage Drop in Consumers' Installations) must be ensured.

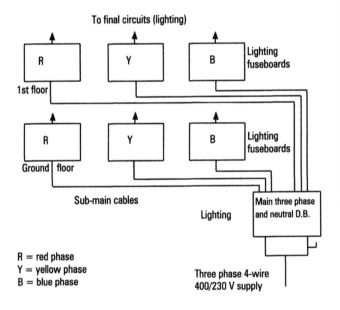

Figure 4.13   Load balancing of lighting loads over the three phases

A further benefit of distributing fluorescent lighting circuits over the three phases in areas where rotating machines are used, is the reduction of the stroboscopic effect.

> *Remember*
> Stroboscopic effect was discussed in Chapter 1.
>
> Go back and refresh your memory if you feel it is necessary.

Extra low voltage tungsten halogen lamp loads

The 12 volt tungsten halogen lamps that were described in Chapter 1, take a much larger current than 230 volt lamps of the same power consumption. With such a small operating voltage, correct lamp voltage is a very important factor for maintaining maximum lumen output. A slight reduction of 3 volts in the voltage supplied to a 230 volt lamp would be fairly insignificant, but a similar voltage reduction to a 12 volt lamp would produce a considerable drop in lumen output.

Where a remote transformer is used to supply a 12 volt lighting load, such as track lighting for display purposes, a long length of cable may be required. Consequently, allowance must be made for the voltage drop in the cables between the load and the transformer.

Conventional double-wound transformers produce a drop in output voltage as the load on the transformer increases. The graph in Figure 4.14 shows the reduction in output voltage as the power of the lighting load is increased. If the transformer is not carrying its full load, its output may slightly exceed 12 V. Minimum lamp load is sometimes stated in the manufacturers' data.

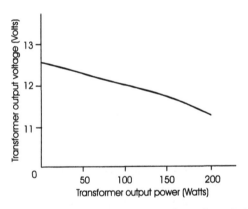

Figure 4.14   Load characteristics on a low voltage lighting transformer

Maximum cable lengths

Example Four
The following example describes the procedure to determine the maximum length of cable between a remote transformer and a 12 volt lighting load.

A 200 watt 12 V double wound transformer is used to supply six separate 20 W, 12 V tungsten halogen lamps as shown in Figure 4.15.

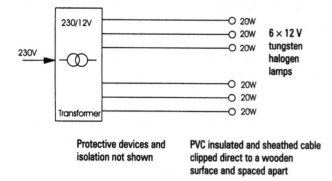

Figure 4.15   Wiring to a lighting transformer

It is recommended that the voltage at a tungsten halogen lamp should not drop below 11.4 V.

Procedure

Total load on transformer secondary
$= 6 \times 20 = 120$ W

From the graph in Figure 4.14, the transformer output voltage has dropped to approximately 11.9 V when a load of 120 W is connected to its secondary terminals.

Maximum voltage drop in each cable
$= 11.9 - 11.4$
$= 0.5$ V

Current in each cable
$= \dfrac{P}{V} = \dfrac{20}{12} = 1.666$ A

Assume a twin PVC insulated cable with copper conductors is used for each lamp circuit (Appendix 4, Tables 4D 2A and 4D 2B of BS7671) and no derating factors are required.

If a 1 mm² cable is selected, the voltage drop is 44 mV/A/m, therefore:

Maximum length of cable run to each lamp

$= \dfrac{\text{Maximum voltage drop} \times 1000}{\text{mV/A/m} \times \text{lamp current}}$

$= \dfrac{0.5 \times 1000}{44 \times 1.666}$

$= 6.82$ metres

For a longer length of cable run, a larger cross sectional area of conductor would be selected, and the calculation repeated to determine its maximum length.

> *Try this*
> A short length of 12 V lighting track contains a total load of 150 W of tungsten halogen lamps. The track is supplied by a 230 V/12 V double wound transformer.
>
> If 1.5 mm² two core PVC insulated and sheathed cable is used to supply the lighting track, determine the maximum distance that the transformers can be located away from the track. Terminal voltage of transformer on a load of 150 watts is 11.7 V. Lamp voltage should not drop below 11.4 V.
>
> Current-carrying capacity of 1.5 mm² cable = 19.5 A.
> Voltage drop (per ampere per metre) = 29 mV
> (No derating factors are required)

High voltage discharge signs

The translucent tube of a high voltage discharge sign (for example neon sign) is hermetically sealed and designed for the emission of light that arises from the electric current passing through the vapour or gas which is contained within it. The tube may be with or without a fluorescent coating on its inner surface.

Figure 4.16 shows a typical high voltage neon sign installation.

The Requirements for Electrical Installations (IEE Wiring Regulations) BS 7671 stipulate the requirements for isolation and switching of the low voltage supply to the neon sign.

> *Remember*
>
> Isolate the H.V. supply when working near neon signs.

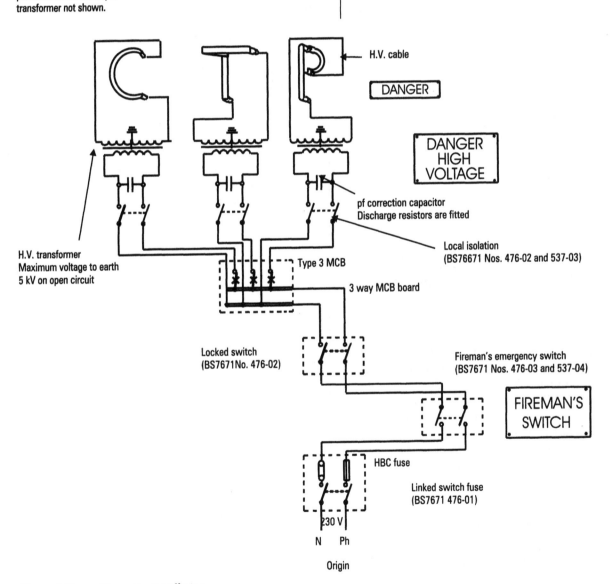

Figure 4.16   Neon sign installation

Fireman's emergency switch

The purpose of the fireman's switch is to isolate the high-voltage discharge sign in the event of a fire occurring in a building, and therefore reducing the danger to the firemen when they are playing their hoses onto the equipment and building.

Fireman's switches are connected in the low voltage circuits supplying exterior or interior discharge lighting installations operating at a voltage exceeding low voltage for example H.V. neon signs.

For an exterior H.V. sign the switch should be located in a conspicuous position outside the building and adjacent to the equipment, or a notice can be displayed adjacent to the equipment indicating its position. For interior installations, the fireman's switch should be placed in the main entrance of the premises, or in another suitable position to be agreed with the fire authority.

The height of the switch from the ground must not exceed 2.75 metres, as shown in Figure 4.17.

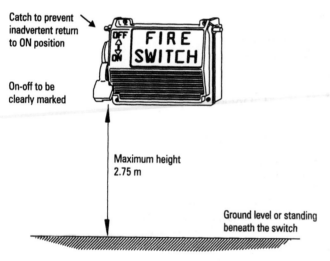

Figure 4.17

Fireman's switches are painted red, with ON and OFF positions clearly indicated, and means to prevent inadvertent return to the ON position. OFF position is at the top of the switch (that is up for OFF) so that the switch can be pushed to this position with a pole. A plate should be placed adjacent to the switch and marked with the words "FIREMAN'S SWITCH", the lettering easily legible. The plate should be a minimum size of 150 mm by 100 mm as shown in Figure 4.18.

Figure 4.18

The Locked switch shown in Figure 4.16 is a special form of switch that can isolate the supply to the H.V. sign and equipment with a lock and key. A switch with a removable handle when switched off can also be used for the same purpose. The key or handle must not be interchangeable with any other type of switch which may be used for a similar purpose in the building. The electrician can keep the key (or handle) in his possession while working in safety, near or on the sign.

L.V. circuit protection

Extra protection against earth faults on the low voltage circuits can be achieved by connecting a residual current operated circuit breaker at the origin of the supply.

Try this

1. The load current of a number of fluorescent luminaires is required but the power factor is unknown. State the formula that could be used to find the current.

2. At which position is the handle of the FIREMAN'S SWITCH when the H.V. sign is operating under normal conditions?

3. Determine the current taken by 10 × 75 W single lamp 230 V fluorescent luminaires.

Emergency lighting

Figure 4.19   Emergency luminaire

United Kingdom laws and standards

An emergency lighting system may be required when the main lighting fails. There is considerable legislation and standards in operation which cover the requirements for different types of premises where emergency lighting is needed. For example, legislation in the form of the Cinematograph Regulations 1955 lays down the rules that govern the emergency lighting provision in cinemas. The Fire Precautions Act, 1971 and the Health and Safety at Work Act 1974 make it obligatory to provide means of escape in all places of work and public resort. The provision of an emergency lighting system is an essential part of this requirement. Legislation in the form of the Fire Precautions (Workplace) Regulations 1997 is considerably more wide-ranging than the 1971 Fire Precautions Act. British Standard 5266:Pt 7, (European Standard EN1838): 1999, Lighting applications – Emergency lighting, provides details of the minimum standards required for design purposes.

It is important therefore that during the design stages of an emergency lighting system a considerable amount of information is required that covers all the current British Standards and legislative requirements applicable to the premises. The following notes provide a brief insight into the requirements of an emergency lighting system.

European Community requirements

UK Laws and Standards applied to the provision of emergency lighting are well established, but as the UK and the other members of the European Community are in the process of harmonisation of standards, fundamental changes in the provision for emergency lighting are taking place.

The notes in this part of the studybook include references to the UK Standards and the European requirements for emergency lighting.

Escape lighting

This type of emergency lighting should ensure that it is possible for the premises to be evacuated safely and effectively. It must:
(a) indicate clearly and unambiguously, all the escape routes in the premises
(b) provide sufficient illumination for all the escape routes to enable safe movement of personnel towards and out of the exits in the premises.
(c) ensure that all fire equipment and fire alarm call points which are provided along the escape routes in the premises can be readily located by personnel

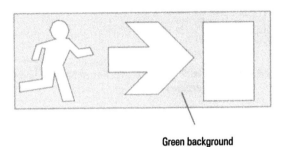

Figure 4.20

Design procedure

All the relevant information for the emergency lighting project should be sought and gathered together.

This should include:

>   legislative requirements (national and/or local)

>   customer preferences in consultation with architect, design engineer, installation contractor, enforcing authority (for example fire authority) and any other interested party

>   type of system, for example, central battery and remote slave luminaires or self-contained luminaire in which each luminaire has its own battery.

Initial design of the emergency lighting system will be concerned with locating the luminaires at the points of emphasis so that they cover specific hazards and highlight safety equipment and signs. This procedure is performed regardless of whether the particular area is an emergency escape route or an open area. When all the essential locations (points of emphasis) have been covered, the type of luminaire required and its light output can be considered. In addition to these luminaires, it may be necessary to instal extra luminaires for escape route lighting to ensure that the minimum illuminance levels are achieved.

The positioning of **essential escape lighting** is as follows (Figures 4.21 (a)–(j)):

   Denotes emergency lighting luminaire

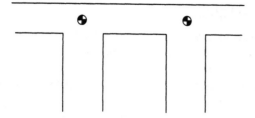

Figure 4.21(a) Positioned at each intersection of corridors

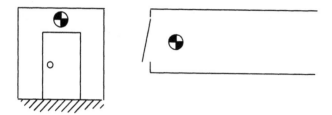

Figure 4.21(b) At each exit door which is to be used in an emergency

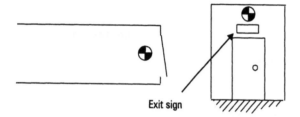

Figure 4.21(c) To illuminate mandatory emergency exit and safety signs

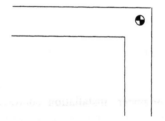

Figure 4.21(d) At each change of direction

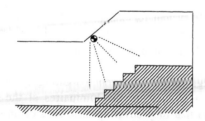

Figure 4.21(e) Near each staircase so each flight receives direct light

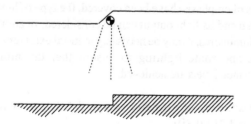

Figure 4.21(f) At each other change of floor level

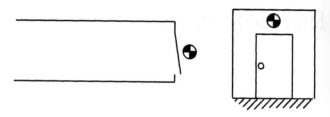

Figure 4.21(g) Outside each final exit and close to it

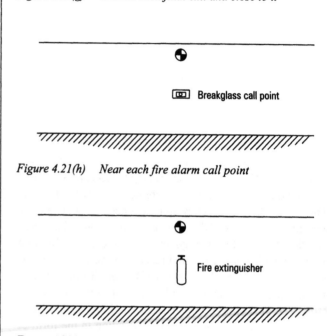

Figure 4.21(h) Near each fire alarm call point

Figure 4.21(i) Near fire fighting equipment

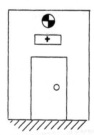

Figure 4.21(j) Near each First Aid point

Note: "Near" is considered to be within 2 metres measured horizontally.

BS 5266-7/EN1838 requires an illuminated directional sign (or a series of signs) to be installed if it is not possible to have direct sight of the emergency exit. The sign will assist personnel to progress in the direction of the emergency exit.

Having positioned the essential escape lighting luminaires, consideration must also be given to any other area requiring emergency escape lighting cover.

xamples include:
- open areas (for example auditoria)
- covered car parks
- external to the premises to provide safe evacuation
- moving stairways, walkways, lifts
- toilets, lobbies, closets
- high risk areas such as those containing rotating machinery
- switchgear control rooms

Levels of illuminance

A minimum illuminance level of 1 lux (BS 5266-7, European Standard EN 1838:1999) is required on the floor along the **centre line** of clearly defined **ESCAPE ROUTES** that are up to 2 m wide. Also, across the central band, consisting of not less than half the width of the escape route, is required to be illuminated to a minimum of 50% of the stated value. A UK deviation (at the time of writing) to the above value is 0.2 lux along the centre line for escape routes that are **permanently unobstructed**. However, it is very difficult to guarantee that escape routes remain completely unobstructed at all times consequently the minimum value of 1 lux is strongly recommended for new installations.

If the mains lighting fails, the **escape route** emergency lighting must reach 50% of the required level of illuminance within 5 seconds and full emergency illuminance within 60 seconds.

If the positions shown in Figures 4.21(h), 4.21(i) or 4.21(j) are not on the escape route nor in an open area, a minimum illuminance of 5 lux is required at floor level.

Consideration should also be given to the possibility of emergency luminaire failure in any "lighting compartment" (for example isolated corridor section) of an escape route. In this situation a minimum of two luminaires should be installed to ensure that the compartment of the escape route is not plunged into darkness.

Note: wider escape routes may be treated as a series of 2 m wide bands or make provision for open-area (anti-panic) emergency lighting.

Open area (anti-panic) emergency lighting

For areas larger than 60 m^2, or open areas with undefined escape routes, emergency lighting and signs should be installed.

Anti-panic emergency lighting is provided to avoid panic by having sufficient illumination to allow people to reach an identified escape route. A minimum emergency illuminance of 0.5 lux is required at floor level in the empty core area, excluding the 0.5 m border of the perimeter of the area.

In the event of mains lighting failure, the illuminance in the anti-panic area must reach 50% of the required level of illuminance within 5 seconds, and full emergency illuminance within 60 seconds.

Try this

1. What is meant by the term "Slave luminaire"?

2. Which Code of Practice should be referred to during the design of an emergency lighting system?

3. State two functions of escape lighting.

Remember
An emergency lighting system is a legal requirement for many public, commercial and industrial premises.

Remember
Glare must be avoided in emergency lighting installations.

> **Try this**
> State briefly the type of information required during the design of an emergency lighting scheme.
>
>

The method of calculating the number of luminaires required is the same as described in Chapter 3 – Lighting Design. However, a further factor K for the reduction in battery voltage or cable volt drop or both, where applicable, is used in the formula to determine the number of luminaires.

Example Five
An open area requires an illuminance of 1 lux at floor level from fluorescent emergency luminaires.

Empty core area dimensions are 30 m × 17 m (these dimensions exclude the 0.5 m to the perimeter of the empty core area).

Assume a maintenance factor (MF) of 0.8.

Photometric data for the emergency luminaire show that:

Utilisation factor (UF) = 0.28 (with ceiling, wall and floor reflectances at zero)
Emergency lighting lumens (ELL) = 360 lm
Factor K to allow for battery discharge = 0.85

The total number of emergency lighting luminaires must be determined to provide an illuminance of 1 lux (E).

Number of luminaires

$$= \frac{E \times \text{Room dimensions } L \times W}{ELL \times UF \times MF \times K}$$

$$= \frac{1 \times 30 \times 17}{360 \times 0.28 \times 0.8 \times 0.85}$$

= 7.44 luminaires

Actual number will be 8 luminaires.

The spacing of the luminaires will depend upon their mounting height, and also the maximum space-height ratio from the photometric data of the manufacturer.

In HIGH RISK TASK AREAS that are likely to contain potential hazards, such as rotating machines, conveyors and acid baths, emergency lighting equivalent to not less than 10% of the normal illuminance of mains lighting levels must be provided on the reference plane of the respective task. However, the level of illuminance must not be less than 15 lux. The emergency lighting must be free of stroboscopic effects. Also, the required level of illuminance shall be provided within 0.5 seconds and continue for as long as the particular hazard exists.

Battery systems
There are two types of battery systems of supply: CENTRAL SYSTEMS, where a bank of batteries is remote from the emergency lighting luminaires; and SELF-CONTAINED SYSTEMS where each luminaire has its own battery.

Modes of operation
There are three basic modes of operation;

MAINTAINED, NON-MAINTAINED and SUSTAINED.

MAINTAINED emergency lighting is where the lamps of the luminaire operate from the normal mains supply or the emergency battery supply when the mains supply fails

NON-MAINTAINED emergency lighting is where the emergency lamps are only in operation when the normal mains lighting supply fails. They operate from the emergency supply.

SUSTAINED emergency lighting is where a luminaire contains two or more lamps and at least one of which is supplied from the emergency lighting supply (battery) and the others from the normal mains lighting supply. If the mains supply fails, the battery supply operates the emergency lighting lamp (or lamps).

The sustained luminaire is often used for exit signs.

Figure 4.22 shows a typical arrangement of self contained (non-maintained) and sustained emergency lighting luminaires in an open area. The mains supply lighting is not shown.

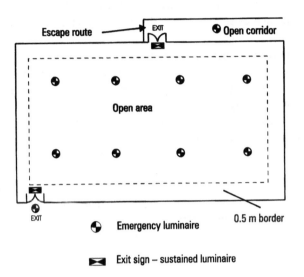

Figure 4.22

The sign shown in Figure 4.23 is in the form of a pictogram, together with the relevant wording that indicates the nature of the sign.

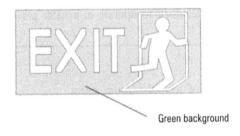

Figure 4.23

Pictogram signs are based on an international standard. The design of the sign ensures that people of all nationalities can understand its meaning. The same style of sign should be used throughout the premises and not mixed with other styles.

European type pictograms exclude wording and rely on graphic symbols only, as shown in Figure 4.24. To meet present and future needs this new pictogram should be used.

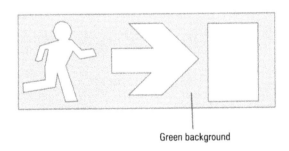

Figure 4.24

Exit signs should be clearly visible on an escape route.

The maximum viewing distance d for an internally illuminated safety sign is $200 \times$ height h of sign. For externally illuminated safety signs the maximum viewing distance d is $100 \times$ height h of sign (Figure 4.25).

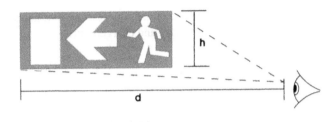

Figure 4.25

Central battery system

Figure 4.26 shows a schematic diagram of a central battery system that provides a maintained supply to signs, and a non-maintained supply to luminaires.

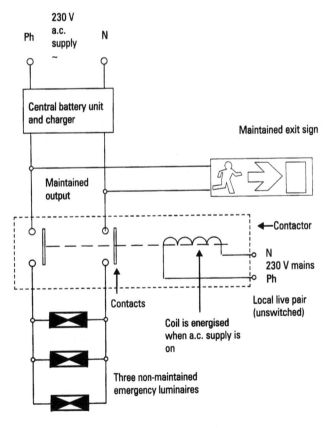

Figure 4.26   Central battery system

65

If the local lighting circuit supply fails (or mains supply fails), the coil will de-energise, and the contactor contacts will close to operate the emergency luminaires.

Section 528 of BS7671 states the precautions to be taken when installing emergency lighting circuits. Reference should be made to this section of BS7671 during design stages of the wiring system.

Self-contained system

Figure 4.27 shows a self-contained fluorescent luminaire which has its own battery supply and charger unit.

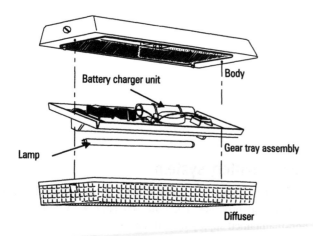

Figure 4.27   Self-contained fluorescent luminaire

Choice of systems

The choice of emergency lighting system, either a **CENTRAL BATTERY SYSTEM** or **SELF-CONTAINED SYSTEM** will depend on the conditions encountered within the various types of premises. However, there are no absolute right or wrong systems, and each system has its merits. Certain premises may use a mixture of both systems.

The merits of a **CENTRAL BATTERY SYSTEM** are:
- maintenance free or longer life batteries are available
- ease of testing and monitoring conditions of cells and equipment at a central point
- ease of testing the operation of all the emergency luminaires in the premises
- fluorescent luminaires which form part of the general lighting of premises can be selected and supplied from a central static inverter and battery, which therefore removes the need for separate emergency luminaires
- particularly suitable for medium to large installations where many emergency luminaires are required.

The merits of a **SELF-CONTAINED SYSTEM** are:
- ease of installation – emergency luminaires can be connected to the nearest unswitched live pair of the lighting circuit
- low maintenance requirement (sealed batteries normally used, for example nickel cadmium)
- ideal for smaller installations (a central system may prove more expensive to install and maintain in smaller premises)
- the system can be easily extended by just adding more luminaires as required to the nearest lighting circuit live pair
- individual final circuit protection (for example if a local lighting circuit protective device operates due to a fault, the self-contained emergency luminaires on that circuit operate immediately).

Emergency lighting conversion kits can also be used to convert the standard fluorescent luminaire into a self-contained emergency luminaire. The lamp operates at its normal light output when the mains supply is on. The conversion kit consists of a battery and inverter which is fitted inside the luminaire. If the mains voltage fails, automatic change-over to the battery and inverter takes place, and the lamp operates at reduced light output.

Although incandescent lamps can be used for emergency lighting schemes, low wattage fluorescent lamps in self-contained luminaires are generally preferred because of their increased light output and low power consumption.

Luminaires that operate for a period between one hour and three hours may satisfy the normal requirements for emergency light. However, many emergency lighting schemes are designed to operate at a safer period of three hours.

Try this

The owner of a six-bedroom guest house requires information on the best and cheapest form of emergency lighting to install in the premises. The guest house is three storeys high and recently decorated throughout. State the type of system you would suggest to the owner, giving reasons for your choice.

Fibre optic emergency lighting

Fibre optic lighting technology is an ideal system for a neat emergency lighting system. It consists of a number of fibre optic cables, and all connected to a central projector box containing tungsten halogen lamps. The box is connected to the emergency power supply (maintained or non-maintained). High intensity light from the tungsten halogen lamps is transmitted by the optical fibres in the cable to the various lighting points in the ceiling, or lower level points as required

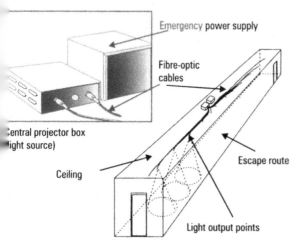

Figure 4.28 Fibre optic emergency lighting system along a corridor

During the installation of the system, the plastic sheathed fibre optic cable must be handled with great care, and the bending radius must not exceed the value stated by the manufacturer.

The merits of the system are:
- since no power or heat is transmitted along the cable, it is ideal for areas where the risk of an explosion exists
- cables are resistant to water and there is no requirement for weatherproof luminaires when illuminating external walkways
- the cables can withstand a temperature as low as –40 °C
- due to the absence of electric fields, the cable can be safely used near radio, TV and communication systems
- only one lamp is required to illuminate a large area
- long life
- ease of maintenance by locating the light source in an accessible position

The standards for the use of fibre optics for emergency lighting applications are shown in BS 5266-4.

Remember

Escape lighting will also be required if a local failure of the normal lighting occurs and the resulting failure produces a hazard, for example, final circuit failure supplying stairway lighting.

Try this

Complete the following:
1. If it is not possible to have direct sight of an emergency exit, an illuminated _____ _____ is required.

2. An emergency lighting luminaire should be positioned at each change of _____ in a corridor.

3. Escape route emergency lighting must reach full illuminance within ____ _____.

4. A high risk task area may contain potential _____.

5. The maximum viewing distance for a safety sign which is _____ illuminated is 200 × height of sign.

6. An emergency exit sign has a _____ coloured background.

7. A self-contained emergency luminaire contains its own _____ and _____ unit.

Standby lighting

In certain situations, such as in hospitals, when the supply to the main lighting fails, it is not always possible for areas to be evacuated immediately. For example, operating theatres may require continuous lighting to the same level of illuminance as prior to the supply failure. In this case the standby supply, probably from a generator, will be required to supply 100% of the design service illuminance of the area. However, the level of standby lighting provided will depend upon the type of activity taking place in the premises and the possible risks that may occur if the supply fails.

Standby lighting is therefore a special form of lighting and must be treated accordingly. Where the standby lighting is for the purpose of emergency escape lighting, the system must comply with the latest BS Code on Emergency Lighting.

Completion of emergency lighting installation

All types of emergency lighting installation must be inspected and tested on completion to the requirements of the latest Regulations for Emergency Lighting.

Regular servicing and testing are also essential to ensure that the system is maintained in correct operation throughout its life.

Air handling luminaires

In large area indoor lighting installations a considerable amount of heat is given off by the lamps and the control gear. Fluorescent type, air handling luminaires are designed to remove this heat before it can enter the room. By careful design of the air flow pattern, air will pass over the lamps and control gear, collecting approximately 70 per cent of the lighting heat which would otherwise enter the occupied area below. The luminaire becomes part of an air conditioning or ventilation system which extracts the stale air from the room. Figure 4.29 shows a section through an air handling luminaire.

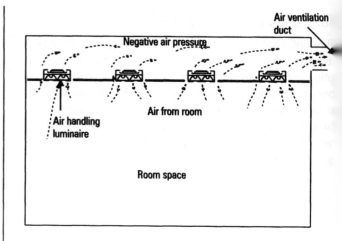

Figure 4.30   Recessed air handling luminaires

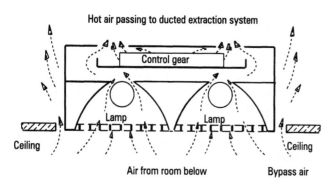

Figure 4.29   Principle of air handling luminaire

The air flow has a cooling effect on the fluorescent tubes which ensures maximum light output, and reduced ceiling temperatures which would otherwise cause discomfort to the occupants below. However, overcooling of certain types of fluorescent lamps can lead to a loss of light output. Restricting the air flow over the lamps by arranging bypass air exhausts outside the luminaire helps to overcome this problem.

Maintaining the correct operating temperature also limits any fine particles of dust on the lamp and luminaire. This leads to reduced maintenance and provides a satisfactory level of illuminance over a longer period of time. The hot air can be extracted via ducts above the luminaires.

Where suspended ceilings are installed, the void formed above the luminaires can be utilised to handle the exhausted air, providing it is fairly air tight and free from any obstructions. The negative plenum formed, as shown in Figure 4.30 can be linked to air ventilation ducts or an air conditioning plant system.

Typical applications

- mainly commercial, for example large area offices and shops.

Remember
Reducing dust deposits on the luminaire helps to improve the value of the maintenance factor for lighting calculations.

Exercises

1. (a) What is meant by the term "a true sine wave"?

   (b) Discuss the problems that may arise when harmonics are present in an electrical distribution system.

   (c) State THREE possible causes of harmonics being present in a distribution system.

2. (a) Discuss the benefits of maintaining a high power factor where a large number of discharge lamps are installed in a factory.

   (b) A lighting load consists of 80, twin-lamp 70 watt fluorescent luminaires, balanced across a three phase, 4 wire 400/230 V supply. Determine the current per phase.

   (c) Describe the allowances that must be made for voltage drop on extra low voltage tungsten halogen lamp loads.

3. (a) Describe the operation of a fireman's switch.

   (b) An electrician is required to work near an H.V. neon sign. What safety precaution should be taken before starting the work?

(c) A large building has three exterior H.V. neon signs fitted. Each sign has its own external fireman's emergency switch.
State the requirements of BS7671 with regard to the marking of these switches.

4. (a) State the reasons for using another factor in the calculation to determine the number of emergency luminaires required for a building.

(b) An office contains six fluorescent luminaires for the general lighting. One of these luminaires is also required for emergency lighting.
Suggest a simple method to adopt in order to make this possible and describe its operation.

(c) State three advantages of using fibre optic lighting for emergency lighting purposes.

(d) What is the function of standby lighting?

(e) Give three reasons why air-handling luminaires may be used in preference to conventional batten type fluorescent luminaires in a large office.

End questions

1. Each outdoor court of a tennis club is to be illuminated by SIX column mounted luminaires containing high pressure sodium lamps.
   (a) Sketch a plan view of the court and indicate the position of each luminaire to provide an even level of illuminance over the playing area.
   (b) Discuss the reason why high pressure sodium lamps are preferred to low pressure sodium lamps for the tennis court.
   (c) If 250 W lamps are used and each lamp has an output of 25 000 lumens, determine the efficacy of the lamps.
   (d) Calculate the line current taken from the three phase, 4-wire 400/230 V supply if all the NINE tennis courts of the club are illuminated in the same manner. The load is balanced across the three phase supply and the power factor is 0.85 lagging.

2. An office measuring 20 m by 20 m is to be illuminated with 35 recessed 70 W twin-lamp fluorescent luminaires. Each lamp has an output of 5700 lumens and the input power to each luminaire is 150 W.
   (a) Determine the average level of illuminance if the maintenance factor is 0.85 and the utilisation factor is 0.59.
   (b) Calculate the line current taken from the three phase 4-wire 400/230 V supply if the power factor is 0.87 lagging and the load is balanced as near as possible across the supply system.

3. (a) Describe the operation of two different types of emergency lighting systems and state the merits of each type.
   (b) Describe the general requirements for hospital ward lighting.
   (c) Discuss the advantages of distributing any type of lighting load across the phases of a three phase 4-wire 400/230 V supply.

4. (a) Exterior car park lighting is provided by FOUR column mounted single-lamp luminaires. A luminaire is fixed at each corner of the 30 m × 30 m parking area and 7 m above the ground. If the luminous intensity of each luminaire is 5000 cd in the direction of the centre area, calculate the level of illuminance at the centre of the car park.
   (b) State the most suitable type of incandescent lamp/luminaire you would select for the car park lighting, giving the reasons for your choice.
   (c) State two types of discharge lamp/luminaire that would also be suitable for the car park.

5. (a) Discuss the various factors that should be considered which will govern the choice of lamp/luminaire for a given indoor lighting installation.
   (b) Describe, with the aid of a diagram, the lamp and luminaire combination you would select to floodlight the front of an office block. What type of control arrangement would you consider to be most suitable?

6. (a) Discuss the advantages of using compact fluorescent lamps in preference to GLS lamps.
   (b) State the recommended level of illuminance (LUX) for the following interiors;
      (i) Education; art rooms
      (ii) Commerce; filing rooms

Answers

The descriptive answers are given for guidance and are not necessarily the only possible solutions.

Chapter 1

p.3 Try this: 1. Improves light output 2. Due to the heating effect of an electric current 3. It has a high melting point 4. Reduced efficacy

p.6 Try this: 1. (a) Choke failure; (b) Lamp failure; (c) Faulty starter switch; (d) No voltage at luminaire (fuse blown or mcb tripped)
2. Lamp will continue to operate. The function of the power factor correction capacitor is to reduce the reactive current taken.

p.8 Try this: 1. The overall power factor of the factory lighting load will be improved. Power consumption of lighting load will be reduced. Reduction in the operating cost of lighting load.
2. The phosphor coating on the inner lamp walls converts the ultraviolet energy produced by the discharge into light.

p.10 Try this: The small quartz enclosure can operate at a much higher temperature than glass.

pp.12 to 14
Exercises
1. Refer to page 2 and pages 4 to 8.
2. (a) Refer to pages 7 and 8; (b) Isolate supply to luminaire and take care that the lamp surface is not contaminated when handling and fitting the lamp.
3. (a) The fault in the glass envelope will not enable the vacuum to be maintained, resulting in a reduction in temperature inside the lamp. Consequently, full vaporisation of the sodium particles will not take place and the lamp will not be able to reach full brilliance; (b) Faulty starter switch (internal short-circuit). The fault will not allow the starting voltage to be produced by the choke. By removing the starter switch when the supply is on, simulates the action of the starter switch to operate the lamp.
4. (a) The MBI lamps have good colour rendering properties and higher efficacy as compared to MBF lamps for this application; (b) Flicker-free starting, extends lamp life, cuts out supply to failed lamps.

Chapter 2

p.18 Try this: 35° = 450 cd, 55° = 275 cd and 65° = 225 cd

p.20 Try this: (a) 2.73 m; (b) A reduction in illuminance on the working plane.

p.22 Try this: 1. Disability glare makes it difficult to see detail. Discomfort glare causes visual discomfort (refer to page 20) 2. Refer to page 21.

p.24 Try this: 1. Space-height ratio 2. (a) L; (b) I; 3. Desk height 4. A square within a square (refer to Table 2.2, page 22)

pp.29 and 30
Exercises
1. approximately 370 cd
2. (a) Refer to page 21; (b) Refer to Table 2.2, page 22
3. (a) Level of illuminance required, type of luminaire, height of luminaires, siting of luminaires; (b) Column mounted luminaires with low pressure sodium lamps (SOX), because of their high luminous efficacy
4 (a) The excess level of illuminance on the working plane can produce a considerable amount of reflected light which may cause visual discomfort and possible fatigue to personnel.
(b) Maintaining the correct level of illuminance and uniformity of the general lighting in this type of premises is important. By adopting a group replacement programme for the lamps will ensure that the general level of illuminance is maintained for a considerable period. Consequently, all items of clothing are highlighted in a uniform manner.
The cost of labour can be substantially reduced by combining lamp replacement with the cleaning of luminaires.

Chapter 3

p.32 Try this: (a) 2.16 m; (b) 787.32 cd

p.33 Try this: P_1 = 133.333 lux, P_2 = 28.80 lux

p.35 Try this: P_1 = 150.4 lux, P_2 = 159.02 lux

p.38 Try this: (a) P_1 = 312.5 lux, P_2 = 72.63 lux;
(b) Floor mounted adjustable SON floodlight luminaire suitable for external use. Economical operating cost, fairly good colour rendering and high lamp efficacy.

p.44 Try this: (a) (i) = 300 lux, (ii) = 12 luminaires for symmetrical layout; (b) 4.37 m

p.45 Try this: 473.76 lux

pp.47 and 48
Exercises
1. P_1 = 29.81 lux, P_2 = 3.066 lux
2. P_1 = 330.57 lux, P_2 = 40.75 lux
3. Illuminance = 500 lux, Luminaires = 18
4. (a) (i) Increased illuminance levels may be required where the room surfaces have poor (dark finish) reflective properties. Light colour surfaces have the opposite effect and may require lower illuminance levels.
(ii) To compensate for the poor reflective surfaces and to provide a satisfactory level of illuminance, more lamps will be required. Consequently, an increase in energy costs. Lower illuminance levels with less lamps will have the opposite effect.

(b) To ensure satisfactory uniformity of illuminance levels on the working plane.
(c) (i) Recessed downlight luminaires with GLS lamps. Illuminance level 300 lux
(ii) High bay luminaires and MBI lamps. Illuminance level 500 lux.

Chapter 4

p.51 Try this: 1. (a) A distorted wave which is the sum of the fundamental and harmonic frequency (or frequencies); (b) Sine wave and complex wave
2. Large harmonic voltage and current

p.53 Try this: Load 1 = 18.04 A, Load 2 = 38.97 A, Load 3 = 11.54 A

p.54 Try this: (a) 179.2 lux; (b) 26 lamps; (c) 6.25 A

p.56 Try this: (a) 50 kW; (b) 52.2 kVA; (c) 16.86°, 0.957 pf lagging; (d) 3.542 kVA$_r$ per phase

p.58 Try this: maximum distance 0.827 m

p.60 Try this:

1. $I = \dfrac{\text{Lamp power (watts)} \times 1.8}{\text{Supply voltage}}$

2. Down position

3. 5.869 A

p.63 Try this: 1. Emergency luminaire (without its own batteries) connected to a central battery supply.
2. Refer to page 61
3. To illuminate staircases and change of direction.

p.64 Try this: Refer to page 61

p.66 Try this: Self contained system (refer to page 66 for reasons)

pp. 69 and 70
Exercises
1. (a) No distortion taking place; (b) Refer to pages 50 and 51; (c) Refer to page 50
2. (a) Refer to page 52; (b) 29.09 A per phase; (c) Refer to page 57.
3. (a) Switch positions: up for off and down for on; (b) Isolate H.V. supply (refer to page 60); (c) Refer to page 60.
4. (a) Refer to page 64.
(b) Refer to page 66.
(c) Refer to page 67.
(d) Refer to page 67.
(e) Refer to page 68.

p.71 End questions
1. (b) The answer should include reference to colour rendering properties of each lamp, efficacy and average life of lamp
(c) efficacy = 100 lumens per watt
(d) line current = 22.92 A
2. (a) 500.24 lux; (b) 8.71 A
3. (a) Refer to page 66; (b) Refer to page 21; (c) Refer to page 57
4. (a) 12.56 lux
(b) Tungsten halogen (refer to pages 4 and 28)
(c) Refer to page 27
5. (a) Refer to page 25
(b) Refer to page 27

6. (a) Refer to page 8
(b) (i) 500 lux; (ii) 300 lux

www.ingramcontent.com/pod-product-compliance
Ingram Content Group UK Ltd.
Pitfield, Milton Keynes, MK11 3LW, UK
UKHW052237090625
459456UK00002B/2